SI: THE INTERNATIONAL SYSTEM OF UNITS

Robert A. Nelson
Byram Hills High School
Armonk, NY 10504

published by

American Association of Physics Teachers
Graduate Physics Building
S.U.N.Y., Stony Brook, NY 11794

May, 1981

PREFACE

This manual is intended as a source of definitions, symbols, conventions, tables, and references for SI units. A brief history of SI is also given. I wish to thank Dr. Lottie E. Mackay of Prentice-Hall Media, Inc., for reading the original manuscript and offering several helpful suggestions. I would also like to thank Ms. Judy Winkler of AAPT for her gracious editorial assistance.

Robert A. Nelson
April, 1980

CONTENTS

I. INTRODUCTION

The modern metric system is known as the International System of Units, _Le Systeme International d'Unites_, abbreviated SI in all languages. The International System of Units is administered by the International Bureau of Weights and Measures (_Bureau International des Poids et Mesures_, BIPM), whose headquarters is the Pavillon de Breteuil in Sevres, France. The BIPM was established by the Treaty of the Meter in 1875, which is now adhered to by 45 countries. The United States was one of the original signatory nations. The Bureau is the repository of the international prototype kilogram and serves to coordinate the work at various national standardizing laboratories. The fundamental scientific decisions are made by the International Committee of Weights and Measures (_Comite International des Poids et Mesures, CIPM_), which meets at least once every two years. It is assisted by the advice of seven consultative committees, each of which deals with one of seven special areas: electricity, photometry and radiometry, thermometry, definition of the meter, definition of the second, ionizing radiations, and units. The decisions of the CIPM are submitted to the General Conference on Weights and Measures (_Conference Generale des Poids et Mesures_, CGPM) for ratification. The CGPM meets every four years. Traditionally, the American representative to the CGPM is the director of the National Bureau of Standards. The work of the International Bureau is reported in the journal _Metrologia_. The day-to-day activities of the Bureau are carried on by a small staff of about fifty persons.[1]

The history of the International System of Units may be divided into four phases which were initiated by the following events: (1) the creation of the original metric system in 1795; (2) the establishment of the International Bureau of Weights and Measures in 1875; (3) the initiation in 1921 of activities designed to redefine the electrical units ; and (4) the extension and revision of the metric system beginning in 1948.

The metric system was the outgrowth of proposals for a uniform, reproducible system of weights and measures set forth in France during the seventeenth and eighteenth centuries. The introduction of this reform was made possible by the French Revolution (1789-1799). Permanence in the standards was sought by basing them on properties observed in nature.

Because French scientists had been especially interested in measurement of the earth's figure, it was natural for them to select as a standard of length a certain fraction of the earth's circumference. The unit expressed by this length, the meter, was defined as one ten millionth of a quadrant of the earth's meridian passing through Paris. The distance was determined from an extrapolation of the measured distance between Dunkirk and Barcelona.[2] All other units were derived from the meter. The unit of volume, the liter, was defined as 0.001 m^3. The unit of mass, the kilogram, was defined as the mass of one liter of water at the temperature of its maximum density (4°C). The metric system of units was officially adopted by the French by decree on April 7, 1795.It was originally known as the Republican system of weights and measures.[3] A decimal calendar was developed that remained in effect for twelve years. The French monetary system was also decimalized.

Other reforms were introduced that, like the metric system, were based on the number 10. The Reaumur temperature scale of 1732 was changed to a centesimal, or centigrade, scale (now called the Celsius scale). For angular measure, the right angle was subdivided into 100 grades.

In 1799 a platinum bar, known as the Metre des Archives, was constructed to serve as the legal representation of the meter. The full length of the bar expressed the value of the unit. A platinum cylinder, the Kilogramme des Archives, was constructed to serve as the legal representation of the kilogram.

By mid-nineteenth century the metric system had attracted international interest, but there was growing concern that the Metre des Archives might not be sufficiently permanent due to abrasion of its ends in use. In 1869 the French government invited representatives of foreign countries to a meeting of an International Commission of the Meter, which was held in Paris the following year. The deliberations were interrupted by the Franco-Prussian War but were resumed in 1872. It was resolved to construct new metric standards from a single ingot of a platinum-iridium alloy and to distribute copies among the participating countries. Also, the International Bureau of Weights and Measures and its governing bodies were created to coordinate the scientific metrological activities of the world. Formal diplomatic approval was provided by the Treaty of the Meter of 1875,[4] which was signed by seventeen countries including the United States. With the establishment of the International Bureau of Weights and Measures the metric system became a truly international system of units. The meter and the kilogram became the ultimate references for length and mass in the United States in 1893.[5]

The further development of the metric system was influenced to a very large extent by advances in the definition of the electrical units. These units were originally devised by physicists and engineers according to their own needs.

The possibility of defining the electrical units in terms of mechanical measurements was first pointed out by Gauss[6] in 1833 and Weber[7] in 1851. Their studies stimulated a complete review of the electrical units by a committee of the British Association for the Advancement of Science in 1861.[8] A fundamental principle stressed by the committee was that a system of units should be coherent. Coherence is the property of one-to-one correspondence between base units and derived units. That is, derived units are expressed as products or quotients of a limited number of base units without requiring numerical factors. The committee selected as base units the meter, gram, and mean solar second. Two parallel systems of units were devised, the electrostatic and electromagnetic subsystems, depending on whether the law of force for electric charges or magnetic poles was taken as fundamental.

The B.A. committee also conducted research on electrical standards. It issued a wire resistance standard, the "B.A. unit," which was intended to represent 10^7 MGS electromagnetic units. This standard soon became known as the "ohm." Later, a standard of capacitance, the "microfarad," was introduced.

In 1873 a second B.A. committee "for the selection and nomenclature of dynamical and electrical units" recommended a change to a centimeter-gram-second system of units[9] because in this sytem the density of water is unity. The CGS unit of force was named the dyne and the CGS unit of work or energy was named the erg. As in the MGS system, any derived electrical unit could be expressed in either electrostatic or electromagnetic measure. For example, the CGS electrostatic unit of charge was the $cm^{\frac{3}{2}} g^{\frac{1}{2}} s^{-1}$ while the CGS electromagnetic unit of charge was the $cm^{\frac{1}{2}} g^{\frac{1}{2}}$. The ratio of the electrostatic to the electromagnetic measure of a given charge was a certain velocity. In his theory of the electromagnetic field Maxwell[10] showed that this velocity was the velocity of light.

The CGS electrostatic and electromagnetic subsystems were combined by Helmholtz[11] and Hertz[12] to obtain the well-known Gaussian system. In addition, still another system of units was introduced by Heaviside[13] and Lorentz[14] to bring about a more satisfying form in the theoretical equations. The system was constructed so as to cause the factors 2π and 4π to be present where they should logically occur on the basis of symmetry. The term which Heaviside used for this modification was "rationalization." Together Hertz and Heaviside were responsible for the modern formulation of Maxwell's equations of electrodynamics.

The recommendations of the B.A. committees were adopted by the First International Electrical Congress in Paris in 1881. Five "practical" electrical units were defined as certain powers of 10 of the CGS units: the ohm, farad, volt, ampere, and coulomb. The second congress (Paris, 1889) added the joule, watt, and a unit of inductance, the quadrant (having the dimensions of length and equal to the length of the earth's quadrant, or 10^9 cm). The quadrant was changed to the henry at the fourth congress (Chicago, 1893). The practical units formed a limited coherent system for electromagnetic quantities but were not coherent with respect to units of mechanical quantities. For example, the ampere was equal to the volt per ohm but it was defined as 0.1 CGS electromagnetic unit, i.e., 0.1 $cm^{\frac{1}{2}}\ g^{\frac{1}{2}}\ s^{-1}$. The joule was equal to the volt ampere second but it was defined as 10^7 g cm^2/s^2 in CGS units.

In 1901 Giorgi[15] demonstrated that the practical electrical units and the mechanical units could be incorporated into a single coherent system by (1) expanding the number of base quantities to four, including one of an electrical nature (he recommended the ohm), and (2) selecting the meter, kilogram, and second as the base units of the mechanical quantities. However, by this time the CGS system was too firmly established for the proposal to be considered seriously.

The situation was complicated in 1908 by the introduction of the so-called "international" electrical units. These units were defined in terms of independent standards of voltage, current, and resistance. Their advantage was that they were conveniently reproducible in a local laboratory. As techniques of measurement improved, however, it was shown that the "international" units were slightly inconsistent. Also, the creation of the national standardizing laboratories made it possible to ship electrical instruments to the testing services for calibration with greater ease and less expense than to perform the calibration experiments first hand.

The International Bureau of Weights and Measures played no active role in the development of the electrical units until 1921, when the Treaty of the Meter was amended to include supervision of the electrical units. Subsequently, in 1929, the International Committee adopted a resolution favoring a return to an absolute system of units, defined in terms of mechanical measurements. The resolution was ratified in

1933 by the Eighth General Conference on Weights and Measures. The Conference delegated to the International Committee the authority to establish the relations between the absolute units and the "international" units and to set the date of their adoption.

A separate question was whether the absolute system of units should be a three-dimensional CGS system or a four-dimensional MKS system, as proposed by Giorgi. Actions were taken on both alternatives. In 1930 the International Electrotechnical Commission (Oslo) redefined or clarified the names maxwell, gauss, oersted, and gilbert, proposed by a committee of the American Institute of Electrical Engineers in 1900, for the CGS electromagnetic units of magnetic flux, magnetic flux density, magnetizing field, and magnetomotive force. Three years later the IEC (Paris) resolved to introduce the permeability of free space μ_0 into the CGS magnetic formulas as the link between electrical and mechanical units. The name weber was recommended for the practical unit of magnetic flux. It was also recommended that the nine practical electrical units should be considered for incorporation into a comprehensive absolute MKS system. After much deliberation this proposal was adopted by the IEC in 1935 (Scheveningen). However, the IEC sought the opinions of the International Committee for Weights and Measures and the Commission for Symbols, Units and Nomenclature (SUN Commission) of the International Union of Pure and Applied Physics as to the choice of the fourth fundamental unit. The International Committee decided that (1) the fourth fundamental unit would be the ampere, defined in terms of the force between two parallel conductors, and that (2) the permeability of free space would have the "rationalized" value $4\pi \times 10^{-7}$ N/A^2. This decision was endorsed in 1938 by a committee of the American Association of Physics Teachers (the committee favored retaining the Gaussian or Heaviside-Lorentz system for use in theoretical work, however.)[16]

The name newton, introduced in 1904 but seldom used, was suggested again in 1935 by L. Hartshorn and P. Vigoureux as the name for the MKS unit of force and was adopted by the IEC in 1938 (Torquay). Finally, in 1946, by the authority given to it by the General Conference in

1933, the International Committee officially adopted the rationalized MKSA system of electrical units to take effect January 1, 1948. The decision was approved retroactively by the Ninth General Conference in October, 1948.[17]

By 1948 the International Bureau of Weights and Measures and its organs were responsible for the units and standards of length, mass, electricity, photometry, temperature, and ionizing radiation. At this time the next major phase in the development of the metric system was begun. It was initiated by a request of the International Union of Pure and Applied Physics, and various other scientific organizations, "to adopt for international use a practical international system of units" based on the MKSA system.[18] Thus the General Conference decided to define a complete list of derived units. Derived units had not been considered previously because they do not require independent standards. Also, the General Conference brought within its province the unit of time, which had been the prerogative of astronomers.

The work was carried on by the Tenth General Conference in 1954 and was virtually completed by the Eleventh General Conference in 1960. During this period there was adopted an extensive revision and simplification of the metric unit definitions, symbols, and terminology. The number of base units was expanded to seven. The meter-kilogram-second-ampere-kelvin-candela system of units was named the "International System of Units (SI)" in 1960. The mole was added as a base unit in 1971.

In advocating the MKSA system the IUPAP specifically did not recommend that the CGS system of units be abandoned by physicists. This approach was assumed to be a compromise between the needs of physicists and engineers. Since then, however, physicists have taken the lead in urging that SI become the one system of units used in all branches of science.

A summary of the historical development of the SI units is given in Table I.

References

1. C. H. Page and P. Vigoureux (eds.), _The International Bureau of Weights and Measures 1875 - 1975_ [Nat. Bur. Stand. (U.S.), Spec. Publ. 420] (U.S. Govt. Printing Office, Washington, 1975), Chap. 1.
2. J. B. Delambre, _Base du Systeme Metrique Decimal_ (Paris, 1806, 1807, and 1810).
3. _Loi Relative aux poids et mesures. Du 18 Germinal, an 3.^{e} de la Republique francaise, une et indivisible_ (Paris, 1795).
4. English translation reprinted in Ref. 1, Appendix 1, p. 225.
5. T. C. Mendenhall, "Fundamental Standards of Length and Mass," Coast and Geodetic Survey Bull. 26 (1893); reprinted in L. V. Judson, _Weights and Measures Standards of the United States: A Brief History_ [Nat. Bur. Stand. (U.S.), Misc. Publ. 247] (U.S. Govt. Printing Office, Washington, 1963), p. 26.
6. K. F. Gauss, _Intensitas vis Magneticae Terrestris ad Mensuram Absolutam Revocata_ (Gottingen, 1833), For extract in English see W. F. Magie, _A Source Book in Physics_ (Harvard Univ. Press, Cambridge, 1963), p. 519
7. W. Weber, "On the Measurement of Electric Resistance According to an Absolute Standard," Poggendorff's Annalen 82, 337 (1851); tr. by E. Atkinson, Philos. Mag. 22, 226, 261 (1861).
8. _Reports of the Committee on Electrical Standards of the British Association for the Advancement of Science_ (reprinted by Cambridge Univ. Press, Cambridge, 1913).
9. J. D. Everett, _Illustrations of the Centimetre-Gramme-Second (C.G.S.) System of Units_ (Taylor and Francis, London, 1875), p. 83.
10. J. C. Maxwell, "On Physical Lines of Force," Philos. Mag. 21 (1861), pp. 161, 281, 338; _ibid_. 23 (1862), pp. 12, 85 [reprinted in _Scientific Papers_ (Cambridge Univ. Press, Cambridge, 1890; Dover, New York, 1965), Vol. I, p. 451]; "A Dynamical Theory of the Electromagnetic Field," Philos. Trans. R. Soc. London 155, 459 (1865) [reprinted in _Scientific Papers_, _op. cit._, Vol. I, p. 526].

11. H. Helmholtz, "Ueber absolute Maasssysteme fur electrische und magnetische Grossen," Wiedemann's Annalen 17, 42 (1882).
12. H. Hertz, "On the Fundamental Equations of Electrodynamics for Bodies at Rest," Wiedemann's Annalen 40, 577 (1890); tr. by D. E. Jones in H. Hertz, Electric Waves (Macmillan, London, 1893; reprinted by Dover, New York, 1962).
13. O. Heaviside, "The Relations Between Magnetic Force and Electric Current," The Electrician, Nov. 18, 1882, p. 6 [reprinted in Electrical Papers (London, 1892; Chelsea, New York, 1970), Vol. I, p. 195].
14. H. A. Lorentz, "Maxwells electromagnetische Theorie," Encyclopadie der Mathematischen Wissenschaften, Band 5, 2. Teil, p. 87 (1904). See also Lectures on Theoretical Physics, tr. by L. Silberstein and A. P. H. Trivelli (Macmillan, London, 1931), p. 69.
15. G. Giorgi, "Unita Razionali di Elettromagnetismo," Atti dell' Associaz. Elettrotecnica Italiana, p. 402 (1901); "Proposals Concerning Electrical and Physical Units," Trans. Int. Electrical Congress St. Louis, I, p. 136 (1904).
16. A.A.P.T. Committee on Electric and Magnetic Units, "What is the Meter-Kilogram-Second System of Units?" Am. Phys. Teach. [Am. J. Phys.] 6, 144 (1938).
17. The International System of Units (SI) [Nat. Bur. Stand. (U.S.), Spec. Publ. 330] (U.S. Govt. Printing Office, Washington, 1977, 3rd ed., p. 17.
18. Ibid, p. 19.

II. THE PROCESS OF MEASUREMENT

The measurement of a physical quantity is the result of a specified sequence of operations. The operations may involve both the use of laboratory equipment and mathematical calculations. Every measurement is a comparison between the physical quantity of interest and some reference quantity of the same character. The specification of a physical quantity thus consists of two parts: its value, represented by a number, and its character, represented by a unit. The value is a measure of the ratio of the quantity determined to the reference of comparison.[1]

A unit of measure is an intrinsically abstract concept. Although the specified operations by which it is realized may be changed, the unit is considered to remain invariant. In contrast, a standard is an arbitrary artifact or reproducible phenomenon chosen as the physical embodiment of the unit. The conditions under which the standard expresses the unit must be specified. For example, when the unit of length is represented by the distance between two lines engraved on a metal bar, the temperature of the bar and the method of support must be specified because of the properties of thermal expansion and elastic deformation.[2]

Units and their standards should have several attributes. First, there should be only one unit for each physically distinct quantity. Second, each unit should be of a size that is convenient to use and should have denominations which facilitate mathematical computation. Third, the standards should be accessible, and to ensure this the operations which define the standards should be reproducible in a sufficiently well-equipped laboratory. Fourth, the operations should be capable of the highest possible precision within the context of accepted theory. Fifth, it would be intellectually satisfying to base each standard on a property of a universal entity, such as the emission frequency of a particular atom, or on a universal constant, such as the speed of light. Sixth, the system of units should be international in character and used in both scientific and everyday affairs. Historically, the standards have included material artifacts, such as the meter bar, and natural phenomena, such as the rotation of the earth. Whatever motives originally guided the choice of definitions, however, all standards used throughout history have been arbitrary in the final analysis.

According to the operational point of view,[3] a measurement of a physical quantity is defined by the operations and calculations of which it is the result. Different people may hold different personal conceptions of a physical quantity, but they must agree on the proper procedure for its measurement. For example, the significance one may prefer to attach to "gravitational mass" might be either "quantity of matter" or "a number defined by Newton's law of gravitation," but the operational definition consistent with the method used to measure it in the laboratory is "the reading on an appropriately calibrated balance."

Every measurement carries with it certain assumptions which require the use of accepted theory. The theory takes our discrete sensory data and systematizes them so as to form a mental picture which we find convenient, if not indispensable. Thus when we measure the mass of an electron we do not follow the usual procedure of placing it on a balance. Rather, we observe pointer readings on our voltmeter and ammeter and make alignments between a ruler and a patch of light in a glass tube. We superimpose on these operations the image of a stream of electrons subject to a force due to a magnetic field and made visible by an interaction with a trace of mercury vapor. We calculate the result using the theory of electrodynamics and call this the mass of an electron.*

Some kind of theory is required for even the simplest measurement. In Newtonian mechanics the measurement of the length of a room, for example, requires the assumption that the transportation of the meter

---

*Of course, this is the inertial mass, defined by Newton's second law.

stick and the location of the room are immaterial to the result. When we say that the distance to the moon is 386 million meters we have in mind that the distance could be obtained by laying a meter stick down end to end 386 million times, even though we would actually obtain the distance by some other method. What can be done in principle, what factors are important to the outcome, and what groups of operations are equivalent are subjects of theoretical physics.

Sometimes an analysis of the distinction between what one actually does and what one thinks one does may lead to a more fundamental understanding of natural phenomena, as in Einstein's analysis of the concept of simultaneity or in Heisenberg's analysis of uncertainty in the measurement of atomic processes. The need for this careful distinction applies in particular to the definition of units. For example, it is theoretically significant to ask whether the unit of time defined by an atomic phenomenon is equivalent to the unit of time defined by the motions of the heavenly bodies.

A group of standards does not by itself constitute a system of units. In establishing a unit system, one must choose the nature and number of metrologically independent quantities. The sizes of the units must also be selected. These decisions are neither obvious nor unique. They depend on the requirements of physical observation and a sense of theoretical elegance.

A fundamental property of a system of units is the property of coherence. Products and quotients of the units should be equivalent to other units within the system without requiring numerical factors. For example, the unit of velocity should be equivalent to the unit of distance divided by the unit of time.

When the metric system was originally created there was only one independent unit, namely the unit of length. The unit of mass was derived from the unit of length by specifying a particular volume and density of water. Later an independent unit of mass was adopted.

The history of the electrical units offers a more complex example. The CGS electrostatic and electromagnetic subsystems, the Gaussian, Heaviside-Lorentz, and practical systems were all different, yet they were all based on the same group of metric standards. The inconvenient size and complicated form of the CGS systems, involving fractional exponents, were indications that the number of base quantities chosen was too small.[4] The difficulties were solved by the Giorgi proposal, which increased the number of base quantities to four. The size of the units also depends on the final form of the equations in which they are used. This property was illustrated by the Heaviside-Lorentz system, which was primarily concerned with mathematical symmetry. Both the Giorgi and the Heaviside-Lorentz innovations were ultimately incorporated into the rationalized MKSA system, and hence into SI.

References

1. A. Gray, Absolute Measuirements in Electricity and Magnetism (Macmillan, London, 1921; reporinted by Dover, New York, 1967), p. 1.
2. Units and Systems of Weights and Measures: Their Origin, Development, and Present Status [Nat. Bur. Stand. (U.S.), Lett. Circ. 1035] (U.S. Dept. of Commerce, Washington, 1976) pp. 1-2.
3. This point of view has been articulated by Eddington. See, for example, A.S. Eddington, The Philosophy of Physical Science (University of Michigan Press, Ann Arbor, 1958). An account of his work may be found in H. Dingle, The Sources of Eddington's Philosophy (Cambridge University Press, New York, 1954).
4. C. H. Page and P. Vigoureux (eds.), The International Bureau of Weights and Measures (1875-1975) [Nat. Bur. Stand. (U.S.), Spec. Publ. 420] (U.S. Govt. Printing Office, Washington, 1975), p. 15.

III. DEFINITIONS OF SI UNITS

In the International System of Units there are seven fundamental units, called base units, and two supplementary units. These are the units of the quantities which are considered to be metrologically independent. The units of all other quantities may be expressed in terms of these units and are called derived units. The base units, supplementary units, and some representative derived units are listed in Table II.

The SI units have the property of coherence. This means that derived units may be expressed as combinations of base units and supplementary units without the need for numerical factors. For example, the m/s is a coherent SI unit of velocity but the km/h is not.

A theory is required to provide the relations between the derived units and the base units. Thus it is a consequence of Newton's second law $\underline{F} = \underline{m}\ \underline{a}$, which relates the net force $\underline{F}$ on a particle to its mass $\underline{m}$ and acceleration $\underline{a}$, that we have the unit relation $N = kg\ m/s^2$. The expressions for the derived units are dictated by the form of the mathematical equations in which they are used. Consequently, the rules for arithmetic used for numbers are applied also to the units.

Multiples or submultiples of the units are expressed by prefixes according to the nomenclature in Table III. This permits the extension of the units over a complete range of sizes without the need for additional names. The use of prefixes should be avoided in calculations, however, and scientific notation should be used instead.

The SI base units at present have the following official definitions (dates in brackets indicate year of adoption by the General Conference on Weights and Measures):[1]

meter (m) [1960]

The <u>meter</u> is the length equal to 1 650 763.73 wavelengths in vacuum of the radiation corresponding to the transition between the levels $2p_{10}$ and $5d_5$ of the krypton-86 atom.

kilogram (kg) [1889; amended 1901]

The kilogram is the unit of mass; it is equal to the mass of the international prototype of the kilogram. [This international prototype made of platinum-iridium is kept at the BIPM under conditions specified by the first CGPM in 1889.]

second (s) [1967]

The second is the duration of 9 192 631 770 periods of the radiation corresponding to the transition between the two hyperfine levels of the ground state of the cesium-133 atom.

ampere (A) [1948]

The ampere is that constant current which, if maintained in two straight parallel conductors of infinite length, of negligible circular cross section, and placed one meter apart in a vacuum, would produce between these conductors a force equal to 2×10^{-7} newton per meter of length.

kelvin (K) [1954; amended 1967]

The kelvin, unit of thermodynamic temperature, is the fraction 1/273.16 of the thermodynamic temperature of the triple point of water.

candela (cd) [1979]

The candela is the luminous intensity, in a given direction, of a source which emits monochromatic radiation of frequency 540×10^{12} hertz and whose radiant intensity in that direction is 1/683 watt per steradian.

mole (mol) [1971]

The mole is the amount of substance of a system which contains as many elementary entities as there are atoms in 0.012 kilogram of carbon-12. Note: When the mole is used, the elementary entities must be specified and may be atoms, molecules, ions, electrons, other particles, or specified groups of such particles.

The precision with which each of these definitions can be reproduced is listed in Table IV. (The precision quoted for the mole is based on the experimental precision of the Avogadro number.)

Of the seven base units, the kilogram is the only one which is still defined in terms of a material artifact rather than a natural phenomenon. This is because the international prototype kilogram can be compared with other masses with greater precision than the mass of an individual atom can be measured. Consequently, increased precision is being sought through the design of improved balances. An atomic standard of mass is not being contemplated at this time.[2]

The supplementary units were admitted as a new class of units in 1960 by the Eleventh General Conference because it could not decide whether they should be considered to be base units or derived units. At present this class contains only two units, both of which are purely geometrical. They have the following definitions:[3]

radian (rad)

The radian is the plane angle between two radii of a circle which cut off on the circumference an arc equal in length to the radius.

steradian (sr)

The steradian is the solid angle which, having its vertex at the center of a sphere, cuts off an area of the surface of the sphere equal to that of a square with sides of length equal to the radius of the sphere.

The supplementary units may be characterized by the established practice that they are either explicitly stated or are suppressed in the statement of the units of a quantity, depending on the application and the taste of the user. For example, the unit of angular velocity may be expressed as either rad/s or simply s^{-1}. It is probably misleading to say that supplementary units "may be regarded either as base units or as derived units."[4] In seeking to clarify their role, Dr. C. H. Page of the National Bureau of Standards has observed:[5]

> There are a few special quantities whose units, for theoretical work, are never changed. For example, all the formulas yielding trigonometric functions of angle are applicable only when the angle is meaured in radians. . . . In situations in which a unit is never subject to change, its dimension can be ignored, and the corresponding quantity treated as if it were simply a number. . . . Units having dimensions which are often conveniently ignored are called "supplementary units" . . . . "Supplementary units" are never "derived units"; they are units whose dimensions are either treated as base dimensions or are ignored.

A specific example which deserves mention is the proper unit for torque. From the theoretical equation for the work done in rotary motion, work = torque x angle, the unit for torque must be N m/rad. But when computing torque from the equation, torque = moment arm x force, we usually express the unit as simply N m. This practice results in confusion with the unit of work and energy, which is also the N m or J. However, the unit of torque[6] is <u>never</u> expressed as J.

To be consistent, the unit for moment arm should be m/rad. In a way this makes sense, because the value of moment arm depends on the point of application and hence the center of rotation. It is therefore not physically equivalent to the distance between two points.

Also, the relative orientations of force and length are different in the computation of work and torque. Following this scheme, the unit for moment of inertia should be kg m^2/rad^2. Then the relation, torque = moment of inertia x angular acceleration, correctly implies that the unit for angular acceleration is rad/s^2. The rule works consistently in other instances as well. Nevertheless, we usually find it mentally convenient to omit the unit "rad" whenever an angular quantity is not explicitly involved. Perhaps we are missing something of theoretical importance by doing this.

The status of the supplementary units within SI has been the subject of reivew by metrologists. A revised approach toward the definition of supoplementary units as "a class of dimensionless, derived units for which the CGPM allows the freedom of using or not using them in expressions for SI derived units" has been considered and approved by the Consultive Committee for Units of the International Committee for Weights and Measures.[7]

References

1. The International System of Units (SI) [Nat. Bur. Stand. (U.S.), Spec. Publ. 330] (U.S. Govt. Printing Office, Washington, 1977), 3rd ed., pp. 3-5.
2. C. H. Page and P. Vigoureux (eds.), The International Bureau of Weights and Measures (1875-1975) [Nat. Bur. Stand. (U.S.), Spec. Publ. 420] (U.S. Govt. Printing Office, Washington, 1975), p. 61.
3. Ref. 1, p. 9.
4. Ibid.
5. C. H. Page, "Classes of units in the SI," Am. J. Phys. 46, 78 (1978).
6. Ibid.; Standard for Metric Practice [ASTM E 380; IEEE Std 268; American National Standard Z210.1] (American National Standards Institute, New York, 1976), p. 7.
7. D. T. Goldman, private communication.

IV. GUIDELINES FOR METRIC PRACTICE

In order to establish uniformity, the International Committee for Weights and Measures (CIPM)[1] and the International Organization for Standardization (ISO)[2] have established guidelines for the use of SI units and of certain other units. The primary working reference for metric practice in the United States is a document[3] prepared jointly by the American Society for Testing and Materials (ASTM) and the Institute of Electrical and Electronics Engineers (IEEE). It agrees with the recommendations of the CIPM and the ISO in all major respects but it differs in the preferred use of some non-SI units and symbols. This document has been approved by the American National Standards Institute (ANSI), which is the American representative to the ISO. Additional U.S. guidelines have been prepared by the IEEE[4] and the National Bureau of Standards (NBS).[5]

A. Symbols

Letter symbols include quantity symbols and unit symbols. Unlike abbreviations, which represent words or names in a particular language, letter symbols represent quantities or units and are independent of language. Symbols for physical quantities are set in italic (sloping) type; symbols for units are set in roman (upright) type (e.g., the value of the force F is 15 N).[6]

A quantity symbol is usually a single letter (e.g., force F, mass m). A unit symbol is denoted by the first letter of the unit name only (e.g., gram is g, not gm; second is s, not sec), with some exceptions (e.g., Hz, Pa, cd, and mol). The unit symbol is not followed by a period. Symbols for unit names derived from proper names have the first letter capitalized but the unit names themselves are not capitalized (e.g., ampere, A). Also, plurals of unit symbols are not followed by an "s" (e.g., 3 kg, not 3 kgs), but plurals of unit names are (e.g., 3 kilograms).

Symbols for prefixes are set with no space between the prefix and the unit. Compound prefixes are to be avoided (e.g., pF, not μμF). There is always a space between the number and the unit symbol following it (e.g., 5 cm, not 5cm), except when the first character of the symbol is not a letter (e.g., 20°C; 30° 15' 6"). A hyphen may be inserted between the number and unit when the number is used as an adjective (e.g., 35-mm film; but the width of the film is 35 mm).[7]

Multiplication of units is indicated by inserting a raised dot or by leaving a space (e.g, N•m or N m, but not N-m or N x m). When the dot is omitted the order of the units in a product should be arranged so as to avoid confusion with other symbols (e.g., N m, not m N, which could be interpreted as millinewton). Division may be indicated by the use of the solidus (/), horizontal fraction bar, or negative exponent (e.g., m/s , $\frac{m}{s}$, or m s^{-1}), but repeated use of the solidus is not permitted (e.g., m/s^2, not m/s/s). The accepted convention is that all symbols apearing to the right of the solidus are interpreted as belonging to the denominator (e.g., J/kg K means $\frac{J}{kg\ K}$).[8] However, to avoid possible misinterpretation the preferred practice is to use parentheses or negative exponents (e.g., J/(kg K) or J kg^{-1} K^{-1}).[9]

Powers of units are indicated by an exponent (e.g., m^2 or cm^3, but not sq m or cc). The exponent applies to the whole unit including its prefix (e.g., cm^3 means $(cm)^3$, not 10^{-2} m^3).

Three-digit groups in multi-digit numbers are separated by spaces instead of commas (e.g., 86 400, not 86,400), but a space is not needed in a four-digit number (e.g., 2000) unless it is part of a vertical column of larger numbers. This convention ensures that a comma will not be mistaken for a decimal marker. The comma is used as the decimal marker in European literature and it is thought that ultimately this practice will become universal. The spacing convention is also used to the <u>right</u> of the decimal marker (e.g., 35 728.123 567).[10]

A zero should be placed in front of the decimal marker in decimal fractions (e.g., 0.3 kg, not .3 kg). Whenever possible the prefix of a unit should be chosen such that the value of the prefixed unit falls between 0.1 and 1000 (e.g., 200 kN, but not 200 000 N). Exceptions are cases where several quantities of the same type but of widely different magnitudes are discussed in the same context (e.g., the forces involved in a particular experiment were 0.05 N, 10 N, 250 N, and 5000 N).

The word "degree" and its symbol (°) are omitted from the unit of thermodynamic temperature expressed in kelvins. However, it is retained in the expression of Celsius temperature (°C). The name "Celsius" was itself a compromise between the terms "centigrade" and "centesimal" formerly used in the United States and Europe, respectively. The preferred practice is to not leave a space between the number and the symbol for Celsius temperature (e.g., 20°C is preferred, 20 °C is acceptable, but 20° C is not acceptable.)[11]

Although the liter is not an official SI unit, it is permitted for use with SI. The lower case "l" and the capital "L" are alternative symbols for the liter. The capital "L" was approved by the CIPM in 1978 and was ratified by the CGPM in 1979.[12] The preferred use in the U.S.[13] is the capital "L" so as to avoid confusing the letter "l" with the number"1".

The kilogram is the only base unit whose name, through historical accident, contains a prefix. Multiples or submultiples are formed by attaching prefixes to the word "gram". This is the only instance of an unprefixed unit that is not a base, supplementary, or derived unit.

The accent in the pronunciation of a prefix is on the first syllable. Thus we say kil' o meter, not kil o' meter. A mi' cro meter is a unit of length, but a mi cro' meter is an instrument for the measurement of length.

The spellings <u>meter</u> and <u>liter</u> are preferred by the National Bureau of Standards.[14] Other organizations, such as the American Society for Testing and Materials, prefer <u>metre</u> and <u>litre</u>.[15] The question of spelling, however, is considered by many to be irrelevant to SI. "Metre" is the international <u>English language</u> spelling (in German it is <u>Meter</u>, in Italian it is <u>metro</u>, etc.).

B. Limitations on Units Outside SI

An important function of SI is to discourage the proliferation of unnecessary units. Each physically distinct quantity has one, and only one, clearly defined unit. However, it is recognized that some units outside SI are so well established that their use is to be permitted. Units permitted for use with SI are listed in Table V. When possible, prefixes should be avoided with these units.

The SI unit of time is the second. The minute, hour, or day should be avoided in technical calculations where practical. However, the km/h may be used as a unit of speed in some cases, as in vehicular speed.

Temperature may be expressed in degrees Celsius, obtained from the thermodynamic temperature T in kelvins by the definition of $t_C = T - 273.15$. This implies that the freezing point of water at atmospheric pressure (0°C) is 0.01°C below the triple point where ice, water, and vapor coexist. Temperature intervals have the same numerical values, whether expressed in K or °C.

When the radian is not a convenient unit for plane angle, the degree or decimal fractions of a degree may be used. However, the uses of minutes or seconds of arc should be restricted to astronomy and cartography.[16]

Large areas should be measured in square kilometers (km^2) or square hectometers (hm^2). The hectare (ha) is equivalent to a square hectometer. Its use should be restricted to land or water areas.[17] Some would say it is now completely obsolete.

Volume may be expressed in liters. The liter should be restricted to volumes of liquids or gases. No prefix other than milli (mL) should be used in conjunction with the liter.[18] Volumes outside the range conveniently described in L or mL should be expressed in m^3. The liter is now defined as 0.001 m^3 exactly. However, between 1901 and 1964 it was defined as the volume of 1 kg of water at the temperature of its maximum density (4°C), which according to experiment was equal to 1000.028 cm^3. (This difference was due to a very small discrepancy between the original definition of the kilogram and the platinum cylinder which was constructed in 1799 to represent it. In 1889 the kilogram was defined by a new platinum-iridium cylinder which had virtually the same mass.) Thus to avoid confusion, highly accurate volume measurements should be cited in m^3.[19]

The tonne, defined as 1000 kg or 1 Mg, should be restricted to commercial usage. In the U.S., the NBS has determined that the term "metric ton" is to be used for this quantity.[20]

Some units, listed in Table VI, are permitted temporarily because of existing practice. However, many measurements involving these units can be easily expressed in SI units with the proper prefix. For example, wavelengths in angstroms can be expressed in nanometers by shifting the decimal point one place (1 Å = 0.1 nm). Nuclear cross sections in barns can be expressed in square femtometers (1 b = 100 fm^2). The kilopascal may be used to express atmospheric pressure instead of using millibars or other units (1 mbar = 0.1 kPa). The kilowatt hour should be restricted to the measurement of commercially supplied electrical energy. Eventually this unit should be replaced by the megajoule (1 kW h = 3.6 MJ).[21]

It is also necessary in some specialized fields to recognize certain units whose relations to SI units must be obtained experimentally. These units are listed in Table VII.

The use of CGS units with special names, based on the centimeter, gram, and second, is not recommended. They are listed in Table VIII. These units include the named Gaussian electrical units, unit names formed with the prefix ab- or stat- (e.g., abampere, statvolt), and named four-dimensional CGS electrical units (biot and franklin). Strictly speaking, the Gaussian units cannot be expressed in terms of SI units because the Gaussian system is a three-dimensional system which does not recognize a base electrical unit whereas SI is a four-dimensional system when only electrical phenomena are considered.[22]

All other units outside SI not falling into any of the above categories, such as those in Table IX, are to be "generally deprecated."

The kilogram-force should never be used as a unit of force or weight. The SI unit of force is the newton. Units of small or large forces may be formed by attaching prefixes (e.g., mN or kN).

The millimeter of mercury (mmHg) should be abandoned in favor of the pascal. Mercury barometers can easily be calibrated in kPa. The SI unit pascal clearly distinguishes the unit of pressure from the unit of length, just as the newton clearly distinguishes the unit of force from the unit of mass.

The millimeter of mercury is defined[23] as the pressure of a column of mercury one millimeter high at 0°C, having an assumed density of 13.5951 g/cm^3, at a standard location where the acceleration due to gravity is 980.665 cm/s^2. Thus a pressure expressed in mmHg only represents the height of a <u>fictitious</u> column of mercury, not the true height as read on a barometer. In practice, the reading of a mercury barometer must be corrected to 0°C, not only to account for thermal expansion of the scale, but also because that is the tempera- ture at which the density of mercury has its nominal value. Addi- tional altitude and latitude corrections for gravity are usually ne- glected.

The torr is used by many scientists instead of the mmHg to avoid an association with length, but the two units are not identical. Assuming that mercury behaves as an incompressible fluid and neglecting variation of gravity with height, the pressure of a column of mercury 760 mm high would be in principle 101 325.014 435 4 Pa. In 1954 the General Conference on Weights and Measures rounded this figure off and defined one standard atmosphere as 101 325 Pa exactly. The torr was subsequently defined as exactly 1/760 standard atmosphere by the British Standards Institution in 1958.[24] Thus although the millimeter of mercury and the torr are nominally equivalent and are used synonymously in the literature, they actually differ by (0.014 435 4/101 325) x 100 per cent, or by about one part in 7 million. One of the functions of SI is to eliminate ambiguities such as this.

The International Committee for Weights and Measures recently placed the atmosphere in the category of units generally deprecated. However, the expression "standard atmosphere" will continue to be accepted for the standard pressure 101 325 Pa (i.e., 101.325 kPa).[25]

The joule, rather than the calorie, should be used to measure heat energy. The calorie is nominally defined as the quantity of heat required to raise the temperature of 1 g of water through 1°C. The energy represented by this unit, however, varies according to the temperature of the water. Thus it is necessary to specify the temperatures at the endpoints of the interval, leading to several different definitions.[26] Since 1948 the National Bureau of Standards has used the thermochemical calorie, defined as 1.484 J exactly, in the field of calorimetry as applied by physicists and chemists, if the term calorie has been used.[27] In the field of nutrition the dietary Calorie (capital C) is equal to the kilocalorie. This is another instance in which the proper use of an SI unit removes a possible ambiguity.

The values of some important physical constants are listed in Table X. These values were obtained by a least squares adjustment of recent experimental data. Also, we list conversion factors from U.S. Customary Units to SI units in Table XI.

References

1. The International System of Units (SI) [Nat. Bur. Stand. (U.S.), Spec. Publ. 330] (U.S. Govt. Printing Office, Washington, 1977), 3rd ed.
2. SI Units and Recommendations for the Use of Their Multiples and of Certain Other Units [ISO 1000] (American National Standards Institute, New York, 1973).
3. Standard for Metric Practice [ASTM E 380; IEEE Std 268; American National Standard Z210.1] (American National Standards Institute, New York, 1976).
4. IEEE Standard Letter Symbols for Units of Measurement [IEEE Std 260; American National Standard Y10.19] (American National Standards Institute, New York, 1978).

5. NBS Guidelines for Use of the Metric System [Nat. Bur. Stand. (U.S.), Lett. Circ. 1056] (U.S. Govt. Printing Office, Washington, 1975).
6. Ref. 4, p. 7.
7. Ref. 4, p. 8.
8. Style Manual (American Institute of Physics, New York, 1970), p. 17.
9. Ref. 3, p. 8; Ref. 4, p. 9.
10. Ref. 3, p. 8.
11. Ref. 3, p. 7; Metric Guide for Educational Materials (American National Metric Council, Washington, 1977), p. 4.
12. P. Giacomo, "News from the BIPM," Metrologia 15, 51 (1979); Ibid., 16, 55 (1980).
13. "Interpretation and Modification of the International System of Units for the United States," Federal Register 41, 54018 (December 10, 1976); Ref. 1, p. 11; Ref. 4, p. 15.
14. Ref. 5.
15. Ref. 3, p. 1.
16. Ref. 3, p. 4
17. Ref. 3, p. 5.
18. Ref. 3, p 4.
19. Ref. 1, p. 26.
20. Ref. 3, p. 5.
21. Ref. 3, p. 5.
22. Ref. 1, p. 13.
23. Handbook of Chemistry and Physics (Chemical Rubber Publ. Co., Cleveland, 1960), 42nd ed., p. 3167.
24. H. G. Jerrard and D. B. McNeill, A Dictionary of Scientific Units (Chapman and Hall, London, 1972), 3rd ed., p. 143.
25. Ref. 12.

26. E. A. Mechtly, The International System of Units: Physical Constants and Conversion Factors [NASA SP-7012] (U.S. Govt. Printing Office, Washington, 1973), p. 16.27.
27. D. C. Ginnings, in Experimental Thermodynamics, edited by J. P. McCullough and D. W. Scott (Butterworths, London, 1968), Vol. I, p. 8.

V. MASS AND WEIGHT

In nontechnical literature there has been an inconsistency in the use of the term "weight." In science "weight" has usually meant "gravitational force," but in common usage it has often referred to the property which is correctly called "mass." Noting the ambiguity which has resulted from this practice, the Third General Conference on Weights and Measures in 1901 amended the definition of the kilogram to declare specifically that it is a unit of mass. The Conference reserved the term "weight" to refer to the gravitational force obtained by multiplying the mass of a body by the value of the acceleration due to gravity (g). For this purpose it adopted the standard value of g to be g_0 = 9.806 65 m/s^2, which was assumed to be the value of the actual acceleration due to gravity at sea level and 45° latitude.[1] This number has been used in the definitions of some non-SI units of force and pressure, such as the kilogram-force and the millimeter of mercury.

An advantage of SI is that it draws a clear distinction between mass and force. The kilogram is the unit of <u>mass</u>; the newton is the unit of <u>force</u>. Nevertheless, it remains necessary to clarify the use of the term "weight" within the context of SI.

Among the various proposals three definitions of weight emerge: (1) a nontechnical synonym for mass; (2) a synonym for the force of gravity; (3) the force required to support a body. The definition one prefers depends on how one perceives the word "weight" is actually used and on one's philosophy as to whether the definition should be compatible with the language of scientists or the general public.

In order to simplify the concept, King suggests that weight be defined as the force required to support a body. This force automatically includes the contributions of both the gravitational force and the apparent centrifugal force on the rotating earth. According to King,[2]

> weight is not just another name for the gravitational force exerted on a body by the earth. Instead it is the force exerted by the body on its support, whether that support is a platform on which the body rests or a suspension from which the body hangs.

This is a neat, concise definition which is both technically accurate and simple to understand. It eliminates the need to introduce the term "apparent weight" and is consistent with the term "weightlessness" as applied to bodies in the state of free fall. Objects which are weightless in an orbiting spacecraft are still acted upon by the force of gravitation due to the earth (otherwise they would not be in orbit!). However, their acceleration toward the earth is the same as the acceleration of the spacecraft and thus they do not appear to be attracted to the floor of the spacecraft. King asserts that weight "is a force which has no other name, and is precisely the quantity which is measured directly on our common weighing devices."[3]

The belief that the distinction between mass and weight should be made clear to the public has been expressed by Haney. "The time has come," he says, "to make popular language consistent with scientific language, and this should be a goal of science education." Following King's approach, Haney recommends that [4]

> the general public should be taught to use the term <u>mass</u> whenever the <u>quantity of matter</u> being exchanged or described is of concern, and <u>weight</u> only when the <u>force required to support a body</u> is of concern.

Haney also recommends that the verb "weigh" be replaced with the phrase "measure the mass" unless a force is actually being determined.

In the authoritative <u>Standard for Metric Practice</u> prepared by ANSI, use of the term "weight" is discouraged:[5]

> Considerable confusion exists in the use of the term weight as a quantity to mean either force or mass. In commercial and everyday use, the term weight nearly always means mass. This non-technical use of the term weight in everyday life will probably persist.
>
> In science and technology, the term weight of a body has usually meant the force that, if applied to the body, would give it an acceleration equal to the local acceleration of free fall. The adjective "local" in the phrase "local acceleration of free

fall" has usually meant a location on the surface of the earth. In this context the "local acceleration of free fall" has the symbol g (sometimes referred to as "acceleration of gravity") with observed values of g differing by over 0.5 per cent at various points on the earth's surface. The use of force of gravity (mass times acceleration of gravity) instead of weight with this meaning is recommended.

The AAPT Metric Advisory Committee has recommended the following definition of weight:[6]

> Weight of an object is the gravitational force on it as measured in the reference frame of the observer (for example, with a spring scale which is calibrated in newtons and not locally in kilograms) when the object is at rest in the observer's frame of reference. The same value of the weight is then attributed to the object even when it is in motion with respect to the observer.

The confusion between mass and weight has been fostered by the practice of using the word "pound" to apply to the names of the units of both mass and force in the U.S. Customary System, or British Engineering System. As used in metrology the pound is a unit of mass, but as used in engineering the pound is a unit of force.

There are two versions of the British Engineering System, depending on whether the pound-mass or pound-force is taken as a base unit. In the "absolute" system the fundamental quantities are length, mass, and time, whose units are the foot, pound-mass, and second. Force is a derived quantity whose unit is the poundal. In the "gravitational" system the fundamental quantities are length, force, and time, with units foot, pound-force, and second. Mass is a derived quantity whose unit is the slug. SI is an absolute system but it is also possible to derive a gravitational metric system. The various systems of mechanical units are compared in Table XII.[7]

In the absolute British Engineering System the units of length and mass are defined as (adjusted by international agreement in 1959)[8]

$$1 \text{ foot} = 0.3048 \text{ meter}$$

$$1 \text{ pound-mass} = 0.453\ 592\ 37 \text{ kilogram}$$

exactly. The unit of force is derived from Newton's second law and is required to be coherent. Thus the force which gives an acceleration of 1 ft/s^2 to a mass of 1 pound-mass is 1 poundal. It follows that

$$1 \text{ poundal} = (1 \text{ pound-mass})(1 \text{ ft/s}^2)$$
$$= (0.453\ 592\ 37 \text{ kg})(0.3048 \text{ m/s}^2)$$
$$= 0.138\ 254\ 954\ 376 \text{ N}$$

exactly. (Often such many-digit conversion factors imply a measurement accuracy which has not been achieved in practice.)

In the gravitational British Engineering System the unit of force is defined by the equation for the standard weight of a body

$$w = m\ g_0$$

at some hypothetical location where the acceleration due to gravity has the standard value g_0. Although g_0 nominally represents the actual gravitational acceleration, it is nothing more than a conversion factor. The standard weight of one pound-mass is one pound-force. Therefore,

$$1 \text{ pound-force} = (1 \text{ pound-mass})(g_0 \text{ in ft/s}^2)$$
$$= (0.453\ 592\ 37 \text{ kg})(9.806\ 65 \text{ m/s}^2)$$
$$= 4.448\ 221\ 615\ 260\ 5 \text{ N}$$

exactly. In practice the suffix "force" is often erroneously dropped. The foot, the pound-force, the second, and the pound-mass do not form a coherent system of units. To make the gravitational system coherent it is necessary to introduce a new unit of mass, the slug. Using Newton's second law, we define the slug as the mass which receives an acceleration of 1 ft/s^2 by a force of 1 pound-force. Therefore,

$$1 \text{ slug} = (1 \text{ pound-force})/(1 \text{ ft/s}^2)$$
$$= (4.448\ 221\ 615\ 260\ 5 \text{ N})/(0.3048 \text{ m/s}^2)$$
$$= 14.593\ 902\ 9 \text{ kg}.$$

very nearly. Since the standard acceleration due to gravity is approximately equal to 32.174 ft/s^2, the slug is approximately 32.174 pound-mass and the pound-force is approximately 32.174 poundal.

Engineers have preferred the gravitational version of the British Engineering System because they use forces more often than masses. It is sometimes said that the gravitational system depends on the effects of gravitation, so that the unit of force is variable. This is not the case, however. An appropriately calibrated spring would not be in error because of this variation. Eventually these rather complicated issues may become unimportant as SI is adopted for use in engineering.

References

1. The International System of Units (SI) [Nat. Bur. Stand. (U.S.), Spec. Publ. 330] (U.S. Govt. Printing Office, Washington, 1977), 3rd ed., p. 16.
2. A. L. King, "Weight and Weightlessness," Am. J. Phys. 30, 387 (1962).
3. A. L. King, "More on Weight and Weightlessness," Phys. Teach. 1, 122 (1963).
4. R. E. Haney, "Issues in Teaching Metric (SI)," Sci. Teach. 45 (3), 13 (1978).
5. Standard for Metric Practice [ASTM E 380; IEEE Std. 268; American National Standards Z.210.1] (American National Standards Institute, New York, 1976), p. 6.
6. M. Iona, "Weight and Weightlessness, "Am. J. Phys. 44, 491 (1976).
7. For further discussion, see F. P. Beer and E. R. Johnston, Jr., Vector Mechanics for Engineers (McGraw-Hill, New York, 1972), 2nd ed., Vol. 2, pp. 480-483; G. Shortley and D. Williams, Elements of Physics (Prentice Hall, Englewood Cliffs, NJ, 1965), 4th ed., pp. 69-71; D. Halliday and R. Resnick, Physics (Wiley, New York, 1978), 3rd ed., pp. 81-84.
8. L. V. Judson, Weights and Measures Standards of the United States: A Brief History [Nat. Bur. Stand. (U.S.), Misc. Publ. 247] (U.S. Govt. Printing Office, Washington, 1963), p. 20.

VI. CLASSIFICATION OF SI UNITS

In SI the mole and the candela are classified as base units. The metrological independence of these units, however, has been questioned by many physicists and chemists. In addition, there are some quantities not covered by SI whose units are important in science and engineering. Among these are the neper and bel (units of logarithmic quantities) and the phon and sone (units of acoustic quantities), which have international recognition and should be considered for inclusion in SI. We wish to examine these units critically; we conclude the chapter with a table proposing how SI units might more reasonably be classified.

The mole

The mole is the unit of "amount of substance." The mole is essentially a number, like the dozen. In this respect it is like the radian, which is a particular mode of subdividing a circle. The concept of the mole arises because of the existence of two methods of measuring mass: a set of relative masses on the atomic scale, based on 1/12 the mass of the carbon-12 atom, and a set of absolute masses on the laboratory scale, based on the mass of the prototype kilogram. Astronomers have yet a third method of measuring mass on the astrophysical scale, based on the mass of the sun. The need for the atomic mass unit and the solar mass unit is due to the fact that atomic masses and astronomical masses can be compared with one another with greater precision than they can be compared with the platinum-iridium cylinder which defines the kilogram.

The conversion factor between atomic mass units and grams (or kilograms) is given by Avogadro's number. One gram is equal to an Avogadro number of atomic mass units. The Avogadro number is similar to the "mechanical equivalent of heat." A calorie has a "mechanical equivalent" of joules. Any measurement of Avogadro's number is equivalent to an absolute measurement of the mass of a carbon-12 atom in grams. The value of Avogadro's number depends on the arbitrary number assigned to the reference atom. When in 1961 the reference atom was changed from oxygen-16 to carbon-12 by agreement between the IUPAP and the IUPAC the value of Avogadro's number was changed accordingly.

We have two recommendations for the reclassification of the mole. First, the mole should be a supplementary unit. Like the radian, the mole is dimensionless. The mole exhibits the characteristic property of a supplementary unit in that it is often suppressed. For example, chemists often quote Avogadro's number as a pure number, but in fact it carries the unit mol^{-1}. Chemists often cite molar mass (which they call "gram-atomic weight" or "gram-molecular weight") in g when actually the unit is g/mol.

Second, the fundamental size should be the kilomole, defined as the number of elementary entities equal to the number of atoms in 12 kilograms of carbon-12. This choice would be consistent with the choice of the kilogram as a base unit. More important, it would remove an obstacle for the acceptance of SI by chemists. Chemists prefer to measure concentrations in moles per liter, instead of moles per cubic meter. If the kilomole were selected as the fundamental size, the mol/L would become a coherent SI unit, since 1 mol/L equals 1 $kmol/m^3$. The atomic mass unit (u) would also become a coherent unit since 1 u is dimensionally and numerically equal to 1 g/mol, or 1 kg/kmol. The universal gas constant would have the same value, whether expressed in Pa L/(K mol) or in Pa m^3/(K kmol). The classification of the kilomole as a supplementary unit would appear to be a simple compromise which would satisfy the preferences of physicists and chemists.

The candela

The candela is the unit of luminous intensity. The candela differs from other SI base units in that its definition depends on the introduction of a biological weighting factor.

In the theory of photometry there are four basic quantities. These are luminance (photometric brightness) of an emitting surface $\underline{L}$

(candela per square meter), illuminance (illumination) of a receiving surface $\underline{E}$ (lux = lumen per square meter), luminous intensity (candlepower) $\underline{I}$ (candela), and luminous flux $\underline{\phi}$ (lumen = candela steradian).

The luminous intensity of a source in a given direction is defined as

$$I = L A$$

where A is the projected area of the source perpendicular to the given direction. The luminous flux is obtained by integration over a solid angle. For example, for a spherically symmetric source we have

$$\phi = 4 \pi I$$

The illuminance of a receiving surface at a large distance from the source is

$$E = I \cos \psi / r^2$$

where r is the distance between the receiving surface and the source and ψ is the angle which the normal to the surface makes with the line connecting the surface to the source. This equation contains the two fundamental laws of photometry, namely the inverse square law and the cosine law. In experimental work the intensities of light sources are compared by placing the sources at the proper distances from a given surface so as to make the illuminance due to each equal.

The mathematical laws of photometry are identical with those of radiometry. The study of photometry differs from radiometry in that only a portion of the electromagnetic energy radiated by the source, the portion to which the human eye is sensitive, is included in the measurements and calculations. Photometric quantities are computed from the total radiant energy of the source using a weighting factor $V(\lambda)$, which is a function of wavelength called the "spectral luminous efficiency."

Until 1979 the standard for luminous intensity was a perfect emitter, or blackbody, at the temperature of solidification of platinum. This standard was approximated by a well of fused thoria surrounded by a crucible of platinum maintained at the freezing point (2045 K) by a high frequency induction furnace. (In actual practice, however, this standard was rarely used. The photometric units were maintained by incandescent electric lamps.) According to the Planck theory of blackbody radiation, the intensity of electromagnetic radiation per unit interval of wavelength at a given wavelength λ, emitted in the normal direction by a blackbody having thermodynamic temperature T, is

$$L_{b\lambda} = (2\, c^2\, h/n^2)\, \lambda^{-5}/[\exp(h\, c/n\, k\, \lambda\, T) - 1]$$

(unit $W/m^3 sr$), where n is the index of refraction of the medium, h is Planck's constant, c is the speed of light in a vacuum, and k is Boltzman's constant. The total luminance is obtained by integrating the product of $L_{b\lambda}$ and the weighting function $V(\lambda)$ over all wavelengths. Thus

$$L_b = K_m \int_0^{\infty} V(\lambda)\, L_{b\lambda}\, d\lambda$$

(unit cd/m^2) where K_m is an arbitrary constant, known as the maximum spectral luminous efficacy; its value depends on the choice of units. In practice, the platinum-point blackbody was a standard of luminance rather than luminous intensity. The luminous intensity of a surface of area A is

$$I_b = L_b\, A$$

In order to make the blackbody standard nearly consistent with older standards of luminous intensity, L_b was assigned the value 60 cd/cm^2 for a blackbody at the platinum point. The constant K_m was then determined by the thermodynamic temperature T. If T = 2045 K, as given by the International Practical Temperature Scale of 1968, then by calculation K_m = 670.8 lm/W. However, on the basis of analogue measurements with electrically calibrated radiometers the value that has been adopted[2] is 683 lm/W. The reciprocal of this number is sometimes called the "mechanical equivalent of light." The value of K_m gives the relation between the lumen and the watt at the wavelength 555 nm (in air) where the function $V(\lambda)$ is maximum.

It can be seen that luminous intensity (in cd) is a fraction of the total radiant intensity (in W/sr) of a light source computed according to a standard weighting procedure. In fact, it has been said that photometry amounts to "nothing more than spectrally weighted radiometry."[3] Thus it may be argued that luminous intensity and radiant intensity are of the same character and do not require separate units.

In October, 1979, the Sixteenth General Conference on Weights and Measures redefined the candela as "the luminous intensity, in a given direction, of a source which emits monochromatic radiation of frequency 540 x 10^{12} hertz and whose radiant intensity in that direction is 1/683 watt per steradian."[4] The frequency of the source was specified because it is independent of the index of refraction of the air which, through an oversight, was not considered in previous definitions. This frequency corresponds to a wavelength of 555.016 nm in air of index 1.000 28. By coincidence it is practically at the crossover point where the curves representing photopic and scotopic vision intersect. Thus the definition does not need to specify the state of visual adaptation. The definition means that (1) a primary standard of light is no longer required, and (2) the complex radiation represented by the platinum-point blackbody is replaced by a monochromatic radiation, so that departures of the spectral power distribution of the standard from the theoretical Planck law are eliminated as a source of error. Therefore, since the basic photometric unit has been redefined in terms of the watt, it should not remain as an SI base unit. (Some physicists also believe that the lumen, rather than the candela, should be considered as the fundamental photometric unit.[5] However, the International Committee has considered this issue and has decided, at least for the time being, to retain the candela as fundamental.[6])

Logarithmic quantities and units

In some fields of study, as in electrical engineering and acoustics, there is a need for a unit of logarithmic quantities (gain, attenuation, level, etc.). Two units have been used, the bel (especially its submultiple, decibel) and the neper. They are defined as follows:[6]

The bel (B) is the logarithmic reference quantity which for a ratio of two power quantities corresponds to the ratio 10, and for a ratio of two field quantities corresponds to the ratio $\sqrt{10}$.

The neper (Np) is the logarithmic reference quantity which for a ratio of two field quantities corresponds to the ratio e and for a ratio of two power quantities corresponds to the ratio e^2.

The two units are related by conversion factors[8]

1 neper = 2 log e bel

1 bel = 0.5 ln 10 nepers

For example, the power gain and voltage amplification in an amplifier circuit between points 1 and 2 are defined as

$$G = 10 \log \frac{P_2}{P_1} \text{ dB} = 0.5 \ln \frac{P_2}{P_1} \text{ Np}$$

and

$$A_V = 20 \log \frac{V_2}{V_1} \text{ dB} = \ln \frac{V_2}{V_1} \text{ Np}$$

(These two quantities are equal only when the impedances in which P_2 and P_1 are developed are equal.) Similarly, the power level with respect to a reference power P_0 is defined as

$$L_P = 10 \log \frac{P}{P_0} \text{ dB} = 0.5 \ln \frac{P}{P_0} \text{ Np}$$

The reference power must be specified. The statement that a certain power level is 7 dB above 1 mW may be expressed L_P (re 1 mW) = 7 dB or in condensed form 7 dB (1 mW), especially in presenting data. When the same reference level is used repeatedly it may be omitted.[9]

Units of logarithmic quantities can be embodied in standards. For example, a standard of attenuation in a transmission line might be a certain resistive network; a standard of light absorption might be a certain thickness of glass. A 1-bel standard of attenuation is one for which the output power is 1/10 of the input power. A 1-bel standard of voltage amplification is one for which the output voltage is $\sqrt{10}$ times the input voltage.

In the theory of acoustics two quantities are of special importance. These are sound intensity level

$$L_I = 10 \log \frac{I}{I_0}$$

and sound pressure level

$$L_p = 20 \log \frac{p}{p_0}$$

Sound intensity level and sound pressure level are logarithmic quantities which are expressed in dB. The reference intensity and reference pressure must be explicitly stated. These values are usually $I_0 = 1$ pW/m^2 and $p_0 = 20\ \mu$Pa.[10]

In general L_I and L_p are not equal. For a free progressive plane or spherical sound wave the intensity in the direction of propagation is[11]

$$I = p^2/\rho\, c$$

where p is the effective, or root mean square, pressure ($p = p_{max}/\sqrt{2}$, where p_{max} is the sound pressure amplitude), ρ is the air density, and c is the speed of sound. It follows that for this type of wave[12]

$$L_I = L_p + 10 \log (p_0^2/\rho\, c\, I_0)$$

Thus L_I will equal L_p only when $\rho\, c = p_0^2/I_0 = 400$ Pa s/m, which may be true under certain atmospheric conditions. However, at a temperature of 22°C and a barometric pressure of 100.125 kPa (751 mmHg), the values frequently used in acoustics calculations, we have $\rho\, c = 407$ Pa s/m. Therefore, for this common case the sound intensity level is about 0.1 dB below the sound pressure level.[13] Additional differences would occur for other waveforms or when boundary effects and dissipative effects must be taken into account.

Sound intensity level and sound pressure level are logarithmic quantities which are determined entirely from the physical characteristics of the source and the medium. They are analogous to quantities from radiometry and do not necessarily represent loudness perceived by the human ear.

Logarithmic quantities are dimensionless, since they are calculated from ratios of two like quantities. In practice they are always expressed in either dB or Np. Thus units of logarithmic quantities have the characteristics of supplementary units. The choice of the neper, rather than the decibel, as an SI logarithmic unit would be appealing from the point of view that the natural logarithm has greater theoretical significance than the common logarithm. However, the decibel is more widely used. Page[14] recommends that the neper should be added to SI as a supplementary unit and that the decibel should be retained as a unit which, like the liter, minute, and hour, is "permitted for use with SI."

Physiological quantities and units

Although SI recognizes a unit for light intensity there is no corresponding unit for sound. The units of photometry have received the attention of lighting and electrical engineers for over a century because of the necessity to standardize and regulate the commercial gas and electricity industries. No similar impetus has come from the field of acoustics. From a purely scientific point of view, units for sound deserve a place in SI on the same footing as those for light. The units of light and sound, however, are measures of physiological quantities which depend on a subjective judgement by a human being. They are thus of a different character than other SI units.

The human ear does not respond equally to sounds having equal intensities but different frequencies. For example, a pure tone having pressure level 20 dB (20 μPa) and frequency 1 kHz is plainly audible, but one having the same pressure level at a frequency of 100 Hz cannot be heard.[15] The response of the ear is usually represented in graphical form. However, the perceived loudness level L_N of a pure tone might be expressed analytically by an equation of the form

$$L_N = K\ A(f,I)\ L_p$$

where L_p is the sound presure level, K is a constant, and A(f,I) is an adopted weighting factor which is a function of frequency and intensity. The function A(f,I) is analogous to the function $V(\lambda)$ in the theory of photometry. The loudness level (sound level, noise level) at a point in a sound field is measured with a sound level meter which uses weighting and time integrating circuits according to standard specifications.[16]

Loudness level is not a purely physical quantity, but entails a subjective evaluation. The internationally recognized unit of loudness level is the phon. One phon is the loudness level of a sound wave which is judged by a normal observer under standard listening conditions as being equally as loud as a sound wave having sound pressure level of 1 dB (20 μPa) at 1 kHz.[17]

The apparent loudness of a sound is not proportional to its loudness level. Therefore, a scale of loudness has been developed in which the apparent loudness is proportional to the number of loudness units. The unit of loudness is called the sone.[18] A standard relation between loudness N in sones and loudness level L_N in phons,

$$N = 2^{(L_N - 40)/10}$$

or

$$\log N = 0.1\ (L_N - 40) \log 2 \approx 0.03\ (L_N - 40)$$

is used in practice.[19]

Another type of physiological quantity occurs in the field of radioprotection. The quantity "dose equivalent" is obtained from "absorbed dose" (in grays) by multiplying by a dimensionless factor called the "relative biological effectiveness" (RBE) of the radiation.[20] In order to avoid health hazards caused by confusion of dose equivalent with absorbed dose, the Sixteenth CGPM adopted the sievert (Sv), equal to the joule per kilogram, for the unit of dose equivalent. This unit is dimensionally equivalent to the gray, but applies to the physiological effect of radiation from a radioactive source.

The candela, the phon, and the sievert have in common the property that they are units of quantities which measure the physiological effect of a phenomenon on a human being rather than the physical properties of the phenomenon itself. They are obtained from units of physical quantities through the use of standard weighting procedures. It would seem appropriate, therefore, to recognize a new class of "physiological units" to distinguish units of this type. Physiological units would be neither base, supplementary, nor derived units. Derived units could be formed by combination of base units and supplementary units with physiological units.

Recommendations for the classification of SI units are summarized in the following table. According to this proposal the structure of SI would consist of 5 base units, 4 supplementary units, 3 physiological units, and derived units. Other units could be added to the physiological category if a need could be demonstrated.

Proposed classification of SI non-derived units

Quantity	Unit	Symbol
	Base units	
length	meter	m
mass	kilogram	kg
time	second	s
electric current	ampere	A
temperature	kelvin	K
	Supplementary units	
plane angle	radian	rad
solid angle	steradian	sr
amount of substance	kilomole	kmol
gain, attenuation, level, etc.	neper	Np
	Physiological units	
luminous intensity	candela	cd
loudness level	phon	p
dose equivalent	sievert	Sv

References

1. R. A. Nelson, "On Finding a Middle Ground for SI," J. Chem. Educ. 56, 661 (1979).
2. J. Terrien, "News from the BIPM," Metrologia 14, 89 (1978).
3. W. R. Blevin and B. Steiner, "Redifinition of the Candela and the Lumen," Metrologia 11, 97 (1975).
4. P. Giacomo, "News from the BIPM," Metrologia 16, 55 (1980).
5. Ref. 3.
6. Ref. 2.
7. Letter Symbols to be used in Electrical Technology, Part 3: "Logarithmic Quantities and Units" [IEC Publ. 27-3] (American National Standards Institute, New York, 1974), p. 11.
8. Ibid.
9. Ibid., pp. 19-21.
10. L. L. Beranek, Acoustics (McGraw-Hill, New York, 1954), p. 13.
11. Ibid., p. 12.
12. Ibid., p. 14.
13. Ibid.
14. C. H. Page, "Classes of Units in the SI," Am. J. Phys. 46, 78 (1978).
15. Ref. 10, p. 399.
16. Specification for Sound Level Meters [S1.4] (American National Standards Institute, New York, 1971).
17. Quantities and Units of Acoustics [ISO 31/VII] (American National Standards Institute, New York, 1978), pp. 12-13.
18. Ibid.
19. Expression of the Physical and Subjective Magnitudes of Sound or Noise [ISO R 131] (American National Standards Institute, New York, 1959), p. 6.
20. J. F. Hogerton, The Atomic Energy Deskbook (Reinhold, New York, 1963), p. 431.

Table I. Historical Development of the SI Units

A. Mechanical Units

Quantity	Date	Unit	Definition	Standard
length	30 March 1791		length of one ten millionth of quadrant of earth along meridian passing through Paris	
	1 August 1793	meter	same	
	18 Germinal, an III* (7 April 1795)	meter	same	brass bar having length 443.44 lignes between its ends
	19 Frimaire, an VIII* (10 December 1799)	meter	length of one ten millionth of quadrant of earth along meridian passing through Paris	_Metre des Archives_, a platinum bar having length 443.296 lignes, representing one meter between its ends, kept at Archives de la Republique in Paris

*Date refers to French Republican calendar, which was in effect from October 5, 1793 (retroactive to September 22, 1792) until January 1, 1806.

1889 (1st CGPM)	meter	distance between lines engraved on international prototype meter bar at 0°C	international prototype meter bar, a platinum-iridium bar kept at International Bureau of Weights and Measures in Sevres, France
1927 (7th CGPM)	meter	distance between lines engraved on international prototype meter bar at 0°C when supported by two cylinders 571 mm apart	international prototype meter bar; recommended value of red cadmium wavelength in air is 643.846 96 nm
1960 (11th CGPM)	meter	1 650 763.73 wavelengths in vacuum of red-orange radiation emitted by ^{86}Kr atom	Engelhard krypton lamp operated at triple point of nitrogen
1975 (15th CGPM)	meter		recommended value of He-Ne laser wavelength stabilized to P(7) line, ν_3 band, of methane is 3 392.231 40 nm; recommended value of speed of light in a vacuum is 299 792 458 m/s.

Table I, continued

Quantity	Date	Unit	Definition	Standard
	1979 (CIPM; to be submitted to 17th CGPM in 1983)	meter	distance travelled in time interval of 1/299 792 458 s by plane electromagnetic waves in vacuum	series of stabilized lasers
mass	1792	grave	mass of one cubic decimeter of distilled water at 0°C	
	1 August 1793	grave	same	brass knob weight having mass of 18 841 grains
	18 Germinal, an III (7 April 1795)	gram	mass of one cubic centimeter of distilled water at 0°C	
	19 Frimaire, an VIII (10 December 1799)	kilogram	mass of one cubic decimeter of distilled water at temperature of its maximum density (4°C)	Kilogramme des Archives, a platinum cylinder having mass 18 827.15 grains, representing one kilogram, kept at Archives de la Republique in Paris, France

	1889 (1st CGPM)	kilogram	mass of international prototype kilogram	international prototype kilogram, a platinum-iridium cylinder kept at International Bureau of Weights and Measures in Sevres, France
time		mean solar second	1/86 400 mean solar day	rotation of the earth on its axis
	1960 (11th CGPM)	ephemeris second	1/31 556 925.974 7 of the tropical year 1900	revolution of the earth around the sun, as represented by Newcomb's <u>Tables of the Sun</u>
	1967 (13th CGPM)	second	9 192 631 770 periods of radiation associated with hyperfine transition of ^{133}Cs atom	weighted average of readings of cesium beam clocks operated at various laboratories, determined by the Bureau International de l'Heure, Paris

(Table I, continued) B. Electrical Units

Quantity	Date	Unit	Definition	Standard
resistance	1860	Siemens unit	resistance of column of mercury 1 m x 1 mm^2 at 0°C	resistance of column of mercury 1 m x 1 mm^2 at 0°C
	1863 (1st B.A. Committee Report)	B.A. unit (ohm)	10^7 MGS em units	resistance of column of mercury 104.8 cm x 1 mm^2 at 0°C
	1865 (3rd B.A. Committee Report)	B.A. unit (ohm)	10^7 MGS em units	resistance of coil of platinum-silver wire
	1873 (B.A. Committee on Nomenclature)	ohm	10^9 CGS em units	
	1881 (1st International Electrical Congress, Paris)	ohm	10^9 CGS em units	

	1884 (International Commission)	legal ohm	10^9 CGS em units	**resistance of column of mercury 106 cm x 1 mm^2 at 0°C**
	1893 (4th International Electrical Congress, Chicago	international ohm	10^9 CGS em units	**resistance of column of mercury at 0°C, mass 14.4521 g, length 106.3 cm**
	1908 (Internat. Conference on Electrical Units and Standards, London)	international ohm	resistance of column of mercury at 0°C, mass 14.4521 g, length 106.3 cm	**resistance of coil of manganin wire of various designs**
	1948 (9th CGPM)	ohm	resistance of conductor carrying current 1 A when potential difference is 1 V	**resistance of coil of manganin wire of Thomas design**
electrical potential, emf	1863 (1st B.A. Committee Report)	volt	10^5 MGS em units	**Daniell cell estimated as 1.1 V**
	1873 (B.A. Committee on Nomenclature)	volt	10^8 CGS em units	**Clark cell estimated as 1.457 V at 15°C**

(Table I, continued)

Quantity	Date	Unit	Definition	Standard
	1881 (1st Internat. Electrical Congress, Paris	volt	10^8 CGS em units	
	1893 (4th Internat. Electrical Congress, Chicago	international volt	10^8 CGS em units	Clark cell specified as 1.434 V at 15°C
	1908 (Internat. Conference on Elec. Units and Standards, London)	international volt	potential difference across conductor having resistance 1 int. ohm when current is 1 int. amp.	Weston cell specified as 1.018 4 V at 20°C (revised to 1.018 3 V at 20°C in 1910)
	1948 (9th CGPM)	volt	potential difference across conductor when current is 1 A and power dissipated is 1 W	Weston cell specified as 1.018 64 V at 20°C
	1969 (NBS)	volt		Weston cell specified as 1.018 328 6 V at 20°C

	1972 (NBS)	volt		ac Josephson junction, adopted value of 2e/h is 483 593.420 GHz/V
current	1840 (Weber)		$10\ mm^{\frac{1}{2}}\ mg^{\frac{1}{2}}\ s^{-1}$	tangent galvanometer and earth's field
	1863 (1st B.A. Committee Report)		0.01 MGS em unit	
	1873 (B.A. Committee on Nomenclature)	"weber"	0.1 CGS em unit	
	1881 (1st Internat. Electrical Congress, Paris)	ampere	0.1 CGS em unit	
	1893 (4th Internat. Electrical Congress, Chicago	international ampere	0.1 CGS em unit	current which when passed through solution of $AgNO_3$ deposits silver at rate of 0.001 118 g/s
	1908 (Internat. Conference on Electrical Units and Standards, London)	international ampere	current which when passed through solution of $AgNO_3$ deposits silver at rate of 0.001 118 g/s	comparison with standard resistor coil and standard Weston cell

(Table I, continued)

Quantity	Date	Unit	Definition	Standard
	1948 (9th CGPM)	ampere	current which, if maintained in two straight parallel conductors of infinite length, of negligible circular cross section, and placed 1 m apart in a vacuum, would produce between these conductors a force equal to 2×10^{-7} N/m of length	Rayleigh current balance (primary); comparison with standard resistor coil and standard Weston cell
	1964 (12th CGPM)	ampere		gyromagnetic ratio of proton, recommended value of γ_p' is 26.7513 nHz/T (without diamagnetic correction for H_2O)

(Table I, continued) C. Miscellaneous Units

temperature	1742 (Celsius)	degree	boiling point of water is 0° and freezing point of water is 100° at 25 1/4 Swedish inches of mercury (751.6 mmHg) pressure	Delisle mercury thermometer
	1743 (Christin) 1745 (Linnaeus) 1746 (Strömer)	degree	freezing point of water is 0° and boiling point of water is 100°	mercury in glass thermometer
	1793 (Temporary Commission on Republican Weights and Measures, Paris)	degree	freezing and boiling points of water assigned 0° and 100°, respectively, at 760 mmHg	mercury in glass thermometer
	1854 (Kelvin)	absolute degree	ratio of temperatures of two heat reservoirs is equal to ratio of heats absorbed and rejected by a Carnot engine operating between these two reservoirs; difference between freezing and boiling points of water assigned 100° at 760 mmHg	gas thermometer calibrated by Joule-Kelvin porous plug experiment; $T = t_C + 273.7$

(Table I, continued)

Quantity	Date	Unit	Definition	Standard
	1887 (CIPM)	degree centigrade or centesimal	freezing and boiling points of water assigned 0° and 100°, respectively, at 760 mmHg	scale of constant volume hydrogen thermometer (primary); mercury in glass thermometer
	1927 (7th CGPM)	degree centigrade or centesimal		International Temperature Scale of 1927 (six reproducible equilibrium temperatures with assigned numerical values, specified instruments,* and adopted interpolation formulas; ice point and steam point are fundamental)

*The instruments used to measure temperature are the platinum resistance thermometer, the platinum - 10 per cent platinum/rhodium thermocouple, and the monochromatic optical pyrometer.

1948 (9th CGPM)	degree Celsius		International Temperature Scale of 1948
1954 (10th CGPM)	degree Kelvin	1854 thermodynamic scale; triple point of water chosen as fundamental fixed point, assigned value 273.16°K and 0.01°C	International Temperature Scale of 1948; $T = t_C + 273.15$
1960 (11th CGPM)	degree Kelvin		International Practical Temperature Scale of 1948, text revision; triple point of water is fundamental
1967 (13th CGPM)	kelvin	1/273.16 of the thermodynamic temperature of the triple point of water	
1968 (CIPM)	kelvin		International Practical Temperature Scale of 1968 (12 fixed points)
1975 (15th CGPM)	kelvin		International Practical Temperature Scale of 1968, amended edition of 1975

(Table I, continued)

Quantity	Date	Unit	Definition	Standard
luminous intensity	1860 (London Metropolis Gas Act)	candle		sperm candle burning at rate of 120 ± 6 grains/h
	1884	Violle standard		luminous intensity of 1 cm^2 of a platinum surface at temperature of solidification
	1889 (2nd Internat. Electrical Congress, Paris)	decimal candle	1/20 Violle standard	nominally equivalent to British Parliamentary candle
	1893 (PTR*)	candle		Hefner amyl acetate lamp
	1896 (Internat. Electrotechnical Congress, Geneva	decimal candle		Hefner lamp
	1898 (Metropolitan Gas Referees Notification, London)	candle		Harcourt pentane lamp assigned value of 10 candles

1906 (LCE*)	decimal candle		Carcel colza oil lamp assigned value of 9.6 decimal candles when burning at rate of 42 g/h (France)
1909 (LCE*, NPL*, and NBS*; PTR*)	international candle	name adopted for the decimal candle	carbon filament incandescent lamp (France, G.B., U.S.); Hefner lamp assigned value of 0.9 int. candle (Germany)
1921 (International Commission for Illumination)	international candle		carbon filament incandescent lamp
1937 (LCE*, NPL*, and NBS*)	international candle	blackbody at temperature of solidification of platinum assigned value of 58.9 int. c/cm^2	well of fused thoria surrounded by crucible of platinum maintained at freezing temperature by a high-frequency induction furnace

*__Abbreviations of national laboratories:__

LCE	Laboratoire Central d'Electricite (Paris, France)
NBS	National Bureau of Standards (Washington, D.C., U.S.A.)
NPL	National Physical Laboratory (Teddington, England)
PTR	Physikalisch-Technische Reichsanstalt (Germany)

(Table I, continued)

Quantity	Date	Unit	Definition	Standard
	1946 (CIPM)	new candle	blackbody at temperature of solidification of platinum assigned value of 60 new c/cm^2	
	1948 (9th CGPM)	candela	name adopted for the new candle	
	1967 (13th CGPM)	candela	luminous intensity, in the perpendicular direction, of a surface of 1/600 000 m^2 of a blackbody at the temperature of freezing platinum (2045 K) under a pressure of 101 325 Pa	well of fused thoria surrounded by crucible of platinum maintained at freezing temperature by a high-frequency induction furnace (primary); tungsten ribbon incandescent lamp
	1979 (16th CGPM)	candela	luminous intensity, in a given direction, of a monochromatic source at 540 THz having radiant intensity 1/683 W/sr	electrically calibrated radiometer

amount of substance	1803 (Dalton)	gram atom, gram molecule	amount of substance whose mass in g is numerically equal to its atomic or molecular weight	H = 1
	1814 (Berzelius)			O = 100
	1850 (Stas)			O = 16
	1902 (Ostwald)	mole	name adopted for gram atom or gram molecule (*Principles of Inorganic Chemistry*)	
	1902 (Internat. Commission on Atomic Weights)			O = 16 (amu, chemical scale)
	c. 1936			$^{16}O = 16$ (amu, physical scale)

(Table I, concluded)

Quantity	Date	Unit	Definition	Standard
	1940 (Internat. Commission on Atomic Weights)			1 amu (chem.) = 1.000 275(5) amu (phys.)
	1960 (IUPAP)			$^{12}C = 12$
	1961 (IUPAC)			$^{12}C = 12$ (unified atomic mass unit, u) 1 u = 1.000 317 9 amu (phys.) = 1.000 043 amu (chem.)
	1971 (14th CGPM)	mole	amount of substance of a system which contains as many elementary entities as there are atoms in 0.012 kg of ^{12}C (elementary entities must be specified)	$^{12}C = 12$ Avogadro's number estimated as $6.022\ 169(40) \times 10^{23}$ mol^{-1} (Taylor, Parker and Langenberg, 1969)

Table II. Names and Symbols of SI Units

Quantity	Name of unit	Symbol	Expression in Terms of other SI Units
	Base Units		
length	meter	m	
mass	kilogram	kg	
time	second	s	
electric current	ampere	A	
temperature	kelvin	K	
luminous intensity	candela	cd	
amount of substance	mole	mol	
	Supplementary Units		
plane angle	radian	rad	
solid angle	steradian	sr	
	Derived Units		
area			m^2
volume			m^3
density			kg/m^3
frequency	hertz	Hz	s^{-1}
velocity			m/s
acceleration			m/s^2
angular velocity			rad/s
angular acceleration			rad/s^2
force	newton	N	$kg\ m/s^2$
pressure, stress	pascal	Pa	N/m^2
torque, bending moment			N m/rad or N m
kinematic viscosity			m^2/s
dynamic viscosity			Pa s
work, energy	joule	J	N m
power	watt	W	J/s

Table II (continued)

Quantity	Name of unit	Symbol	Expression in Terms of other SI Units
impulse, momentum			N s
gravitational field			N/kg
charge	coulomb	C	A s
electric potential	volt	V	J/C
resistance	ohm	Ω	V/A
conductance	siemens	S	A/V
inductance	henry	H	V s/A
capacitance	farad	F	A s/V
magnetic flux	weber	Wb	V s
electric field (E)			V/m
magnetic field (B) [magnetic induction, magnetic flux density]	tesla	T	Wb/m^2
electric displacement (D)			C/m^2
magnetizing field (H)			A/m
entropy, heat capacity			J/K
specific heat capacity			J/(kg K)
thermal conductivity			W/(K m)
wave number			m^{-1}
radiant intensity			W/sr
radiance			$W/(m^2\ sr)$
luminous flux	lumen	lm	cd sr
luminance			cd/m^2
illuminance	lux	lx	lm/m^2
molar energy			J/mol
molar entropy, molar heat capacity			J/(mol K)
concentration			mol/m^3

Table II (continued)

Quantity	Name of unit	Symbol	Expression in Terms of other SI Units
radioactive activity	becquerel	Bq	s^{-1}
absorbed dose	gray	Gy	J/kg
dose equivalent	sievert	Sv	J/kg

Table III. Multiples of SI Units

Prefix	Symbol	Factor
exa	E	10^{18}
peta	P	10^{15}
tera	T	10^{12}
giga[1]	G	10^{9}
mega	M	10^{6}
kilo	k	10^{3}
hecto[2]	h	10^{2}
deka[2]	da	10^{1}
deci[2]	d	10^{-1}
centi[2]	c	10^{-2}
milli	m	10^{-3}
micro	µ	10^{-6}
nano	n	10^{-9}
pico	p	10^{-12}
femto	f	10^{-15}
atto	a	10^{-18}

[1]Pronounced jig′ a.

[2]To be avoided where practical according to ANSI Z210.1.

Table IV. Precision of the Base Units

Unit	Precision
meter	4 in 10^9
kilogram	1 in 10^8
second	1 in 10^{13}
ampere	3 in 10^6
kelvin	1 in 10^6 at 273.16 K
candela	3 in 10^3
mole	1 in 10^5

Table V. Units Outside SI Permitted for Limited Use With SI

Quantity	Name of Unit	Symbol	Value in SI Unit
time	minute	min	60 s
	hour	h	3600 s
	day	d	86 400 s
temperature	degree Celsius	°C	$t_C = T - 273.15$
plane angle	degree	°	$(\pi/180)$ rad
volume	liter[1]	L, l	0.001 m^3
mass	tonne (metric ton)	t	1000 kg

[1] The symbol L is preferred in the U.S. according to NBS 330.

Table VI. Units Outside SI Permitted Temporarily for Limited Use

Quantity	Name of Unit	Symbol	Value in SI Unit
length	angstrom[3]	Å	10^{-10} m
	nautical mile[1]	nmi	1852 m
plane angle	minute[4]	'	$(\pi/10\ 800)$ rad
	second[4]	"	$(\pi/648\ 000)$ rad
speed	knot [1 nmi/h][1]	kn	(1852/3600) m/s
area	barn	b	10^{-28} m^2
area	hectare[6]	ha	10^4 m^2
energy	kilowatt hour[2]	kW h	3.6×10^6 J
pressure	bar	bar	10^5 Pa
radioactivity	curie	Ci	3.7×10^{10} Bq
exposure	roentgen	R	2.58×10^{-4} C/kg
absorbed dose	rad	rad, rd	10^{-2} Gy
dose equivalent	rem[5]	rem	10^{-2} Sv

[1]Listed in NBS 330 but not in ANSI Z210.1.

[2]Listed in ANSI Z210.1 but not in NBS 330.

[3]Deprecated in ANSI Z210.1; permitted temporarily in NBS 330.

[4]Permitted for use with SI according to NBS 330; accepted for limited use according to ANSI Z210.1.

[5]Not listed in NBS 330 or ANSI Z210.1; included here because the rem is related to the sievert in the same way that the rad is related to the gray.

[6]Permitted for use with SI in U.S. according to NBS 330; accepted for limited use according to ANSI Z210.1.

Table VII. Units Outside SI Whose Values Are Obtained Experimentally

Name of Unit	Symbol	Approximate Experimental Value
unified atomic mass unit[1]	u	$1.660\ 531(11) \times 10^{-27}$ kg
electron volt[1]	eV	$1.602\ 191\ 7\ (70) \times 10^{-19}$ J
astronomical unit[2]	AU	$1.495\ 979(1) \times 10^{11}$ m
parsec [206 264.806 AU][2]	pc	$3.085\ 678 \times 10^{16}$ m

[1]B. N. Taylor, W. H. Parker, and D. N. Langenberg, Rev. Mod Phys. 41, 375 (1969).

[2]C. W. Allen, *Astrophysical Quantities* (Athlone Press, London, 1973), 3rd. ed., p. 16.

Table VIII. Named CGS Units Whose Use Is Not Recommended

Quantity	Name of Unit	Symbol	Value in SI Unit
energy	erg	erg	10^{-7} J
force	dyne	dyn	10^{-5} N
acceleration	gal[4]	Gal	10^{-2} m/s^2
kinematic viscosity	stokes	St	10^{-4} m^2/s
dynamic viscosity	poise	P	0.1 Pa s
charge	electro-static unit[1,2,3]	esu	(10/c) C
charge	electro-magnetic unit[1,3]	emu	10 C
magnetic flux	maxwell[1]	Mx	10^{-8} Wb
magnetic field (B)	gauss[1]	G	10^{-4} T
magnetizing field (H)	oersted[1]	Oe	$(1000/4\pi)$ A/m
magnetomotive force	gilbert[1,3]	Gb	$(10/4\pi)$ A
charge	franklin[2,3]	Fr	(10/c) C
current	biot[3]	Bi	10 A
luminance	stilb	sb	10^4 cd/m^2
luminance	lambert[3]	L	$(10^4/\pi)$ cd/m^2
illuminance	phot	ph	10^4 lx

[1]Relation with SI unit is a correspondence.
[2]c is the speed of light in cm/s (approx. 3×10^{10}).
[3]Listed in ANSI Z210.1 but not in NBS 330.
[4]Deprecated in ANSI Z210.1; permitted temporarily in NBS 330.

Table IX. Units Outside SI Generally Deprecated

Quantity	Name of Unit	Value in SI Unit
length	micron	10^{-6} m
	X unit	1.002 x 10^{-13} m approx.
	fermi	10^{-15} m
plane angle	grade	(π/200) rad
volume	stere	1 m^3
	lambda	10^{-9} m^3
mass	metric carat	200 mg
	gamma	10^{-9} kg
force	kilogram-force	9.806 65 N
pressure	mmHg	(13.595 1)(9.806 65) Pa
	torr	(101 325/760) Pa
	atmosphere	101 325 Pa
energy	calorie (International Steam Table, "IT")	4.186 8 J
	calorie (mean)	4.190 02 J
	calorie (thermochemical)	4.184 J
	calorie (15°C)	4.185 80 J
	calorie (20°C)	4.181 90 J
	Calorie (kilogram, thermochemical)	4 184 J

Table X. Physical Constants

Numbers in parentheses represent standard deviation uncertainty in last two significant figures. Values quoted are from B. N. Taylor, W. H. Parker, and D. N. Langenberg, Rev. Mod. Phys. 41, 375 (1969).

Quantity	Symbol	Value
velocity of light	c	$2.997\ 925\ 0(10) \times 10^{8}$ m/s
Planck's constant	h	$6.626\ 196(50) \times 10^{-34}$ J s
Avogadro's number	N_0	$6.022\ 169(40) \times 10^{23}$ mol^{-1}
electron charge	e	$1.602\ 191\ 7(70) \times 10^{-19}$ C
electron rest mass	m	$9.109\ 558(54) \times 10^{-31}$ kg
proton rest mass	M_p	$1.672\ 614(11) \times 10^{-27}$ kg
neutron rest mass	M_n	$1.674\ 920(11) \times 10^{-27}$ kg
electron charge to mass ratio	e/m	$1.758\ 802\ 8(54) \times 10^{11}$ C/kg
Faraday consant, $N_0 e$	F	$9.648\ 670(54) \times 10^{4}$ C/mol
Rydberg constant, $m\ e^4 / 8\varepsilon_0^2\ h^3\ c$	R_∞	$1.097\ 373\ 12(11) \times 10^{7}$ m^{-1}
permeability of free space	μ_0	$4\pi \times 10^{-7}$ N/A^2 (by definition)
permittivity of free space, $1/\mu_0\ c^2$	ε_0	$8.854\ 185\ 3(59) \times 10^{-12}$ F/m
electrostatic constant, $1/4\pi\ \varepsilon_0$	k	$8.987\ 554\ 3(60) \times 10^{9}$ N m^2/C^2
universal gas constant	R	$8.314\ 34(35)$ J/(mol K)
Boltzmann's constant, R/N_0	k	$1.380\ 622(59) \times 10^{-23}$ J/K
volume of ideal gas, standard conditions	V_0	$22.413\ 64(94)$ L/mol
Stefan-Boltzmann constant, $2\pi^5\ k^4/15\ h^3\ c^2$	σ	$5.669\ 61(96) \times 10^{-8}$ W K^4/m^2
first radiation constant, $8\pi h\ c$	c_1	$4.992\ 579(38) \times 10^{-24}$ J m
second radiation constant, $h\ c/k$	c_2	$1.438\ 833(61) \times 10^{-2}$ K m
Gravitational constant	G	$6.673\ 2(31) \times 10^{-11}$ N m^2/kg^2

Table XI. Conversion Factors

Numbers marked by an asterisk (*) are exact. Values quoted are from E. A. Mechtly, The International System of Units: Physical Constants and Conversion Factors [NASA SP-7021] (U.S. Govt. Printing Office, Washington, 1973).

To Convert From	To	Multiply By
inch	centimeter	2.54 *
foot	meter	0.3048 *
yard	meter	0.9144 *
mile (U.S. statute) [5280 ft]	kilometer	1.609 344 *
pound-mass	kilogram	0.453 592 37 *
slug	kilogram	14.593 902 9
degree Rankine	kelvin	5/9 *
degree Fahrenheit	degree Celsius	$t_C = (5/9)(t_F - 32)$ *
acre [4840 square yards]	square meter	4 046.856 422 4 *
quart (U.S. liquid)	liter	0.946 352 946 *
gallon (U.S. liquid)[231 in^3]	liter	3.785 411 784 *
bushel (for dry commodities only)[2 150.42 in^3]	cubic meter	0.035 239 070 166 88 *
mile per hour (U.S. statute)	meter per second	0.447 04 *
mile per hour (U.S. statute)	kilometer per hour	1.609 344 *
cps (cycles per second)	hertz	1 *
rpm (revolutions per minute)	radian per second	$\pi/30$ *
pound-force	newton	4.448 221 615 260 5 *
poundal	newton	0.138 254 954 376 *
psia (pound-force per square inch, absolute)	kilopascal	6.894 757 2
foot pound-force	joule	1.355 817 9
British thermal unit (IT)	kilojoule	1.055 056

Table XI. Conversion Factors, concluded

To Convert From	To	Multiply By
megaton (nuclear equivalent of TNT)	petajoule	4.20
horsepower [550 foot pound-force per second]	kilowatt	0.745 699 87
footcandle [lumen per square foot]	lux	10.763 910

Table XII. Comparison of Systems of Mechanical Units with SI

Quantity	British Engineering Systems	Metric Systems	
Absolute:		SI (MKS)	CGS
length	foot (ft)	meter (m)	centimeter (cm)
mass	pound-mass (lbm)	kilogram (kg)	gram (g)
time	second (s)	second (s)	second (s)
force	poundal (pdl)	newton (N)	dyne (dyn)
(F = ma)	(pdl = lbm ft/s^2)	(N = kg m/s^2)	(dyn = g cm/s^2)
Gravitational:			
length	foot (ft)	meter(m)	
force	pound-force (lbf)	kilogram-force (kgf)	
time	second (s)	second (s)	
mass	slug (sl)	metric slug (msl) or TME*	
(m = F/a)	(sl = lbf s^2/ft)	(msl = kgf s^2/m)	

*Technische Mass Einheit (engineering mass unit). Cf. H. G. Jerrard and D. B. McNeill, A Dictionary of Scientific Units (Chapman and Hall, London, 1972), 3rd ed., p. 84.

SELECTED BIBLIOGRAPHY

1. Definitions, Conventions, and Practices

D. V. DeSimone (Director, U.S. Metric Study), Engineering Standards [Nat. Bur. Stand. (U.S.), Spec. Publ. 345-11] (U.S. Govt. Printing Office, Washington, 1971).

H. G. Jerrard and D. B. McNeill, A Dictionary of Scientific Units (Chapman and Hall, London, 1972), 3rd ed.

IEEE Standard Letter Symbols for Units of Measurement [IEEE Std 260; American National Standard Y10.19] (American National Standards Institute, New York, 1978).

E. A. Mechtly, The International System of Units: Physical Constants and Conversion Factors [NASA SP-7012] (U.S. Govt Printing Office, Washington, 1973).

Metric Editorial Guide (American National Metric Council, Washington, 1975), 3rd ed.

Metric Guide for Educational Materials (American National Metric Council, Washington, 1977).

NBS Guidelines for Use of the Metric System [Nat. Bur. Stand. (U.S.), Lett. Circ. 1056] (U.S. Dept. of Commerce, Washington, 1975).

L. D. Pedde et al., Metric Manual (U.S. Govt. Printing Office, Washington, 1978).

SI Units and Recommendations for the Use of Their Multiples and of Certain Other Units [ISO 1000] (American National Standards Institute, New York, 1973).

Symbols, Units, and Nomenclature in Physics [Document U.I.P. 20] (International Union of Pure and Applied Physics, 1978); Physica 93A, 1 (1978).

Standard for Metric Practice [ASTM E 380; IEEE Std 268; American National Standard Z210.1] (American National Standards Institute, New York, 1976).

The International System of Units (SI) [Nat. Bur. Stand. (U.S.), Spec. Publ. 330] (U.S. Govt. Printing Office, Washington, 1977), 3rd ed.

USMA Metric Practice Committee, Metric Units of Measure and Style Guide (U.S. Metric Association, Boulder, 1979).

L. Young, Systems of Units in Electricity and Magnetism (Oliver a Boyd, Edinburgh, 1969).

2. General Histories

A. V. Astin, "Standards of Measurement," Sci. Am. 218 (6), 50 (1968).

E. S. Barr, "Concerning the Naming of the 'Practical' Electrical Units," Am. J. Phys. 29, 532 (1961).

H. Barrell, "The Metre," Contemporary Phys. 3, 415 (1962).

W. K. Burton (ed.), Measuring Systems and Standards Organizations [S. R. 10] (American National Standards Institute, New York, n.d.).

R. C. Cochrane, Measures for Progress: A History of the National Bureau of Standards (U.S. Govt. Printing Office, Washington, 1966).

M. Danloux-Dumesnils, The Metric System: A Critical Study of Its Principles and Practices (Humanities Press, Atlantic Highlands, N.J., 1969).

L. A. Fischer, History of the Standard Weights and Measures of the United States [Nat. Bur. Stand. (U.S.), Misc. Publ. 64] (U.S. Govt. Printing Office, Washington, 1925).

W. Hallock and H. T. Wade, Outlines of the Evolution of Weights and Measures and the Metric System (Macmillan, New York, 1906).

L. V. Judson, Weights and Measures Standards of the United States: A Brief History [Nat. Bur. Stand. (U.S.), Misc. Publ. 247] (U.S. Govt. Printing Office, Washington, 1963).

A. E. Kennelly, "Historical Outline of the Electrical Units," Proc. Soc. Promotion Engr. Educ. 19, 229 (1928).

W. F. Kieffer, The Mole Concept in Chemistry (Van Nostrand, New York, 1973), 2nd ed., Chap. 1.

H. A. Klein, The World of Measurements (Simon and Schuster, New York, 1974).

W. E. Middleton, A History of the Thermometer and Its Uses in Meteorology (Johns Hopkins Press, Baltimore, 1966).

----- The History of the Barometer (Johns Hopkins Press, Baltimore, 1964).

H. Moreau, Le Systeme Metrique: Des anciennes mesures au Systeme International d'Unites (Chiron, Paris, 1975).

-----"The Genesis of the Metric System and the Work of the International Bureau of Weights and Measures," tr. by R. E. Oesper, J. Chem. Educ. 30, 3 (1953).

A. O. Nier, "A New Unified Scale for Atomic Masses and Weights," Phys. Teach. 1, 11 (1963).

C. H. Page and P. Vigoureux (eds.), The International Bureau of Weights and Measures (1875 - 1975) [Nat. Bur. Stand. (U.S.), Spec. Publ. 420] (U.S. Govt. Printing Office, Washington, 1975).

F. B. Silsbee, Establishment and Maintenance of the Electrical Units [Nat. Bur. Stand. (U.S.), Circ. 475] (U.S. Govt. Printing Office, Washington, 1949).

----- Systems of Electrical Units [Nat. Bur. Stand. (U.S.), Monogr. 56] (U.S. Govt Printing Office, Washington, 1962).

C. F. Treat (Historian, U.S. Metric Study), A History of the Metric System Controversy in the United States [Nat. Bur. Stand. (U.S.), Spec. Publ. 345-10] (U.S. Govt. Printing Office, Washington, 1971).

Units and Systems of Weights and Measures: Their Origin, Development, and Present Status [Nat. Bur. Stand. (U.S.), Lett. Circ. 1035] (U.S. Dept. of Commerce, Washington, 1976).

J. W. T. Walsh, Photometry (Constable, London, 1958), 3rd ed., Chap. 1.

E. Wichers, "Can the Scales of Atomic Weights and Nuclidic Masses Be Unified?" Phys. Today 12 (1), 28 (1959).

R. E. Zupko, British Weights and Measures: A History from Antiquity to the Seventeenth Century (Univ. of Wisconsin Press, Madison, 1977).

3 Standards, Measurement, and Calibration

B. E. Blair (ed.), Time and Frequency: Theory and Fundamentals [Nat. Bur. Stand. (U.S.), Monogr. 140] (U.S. Govt. Printing Office, Washington, 1974).

W. K. Clothier, "A Calculable Standard of Capacitance," Metrologia 1, 36 (1965).

J. G. Coffin, "Construction and Calculation of Absolute Standards of Inductance," Bull. Bur. Stand. (U.S.) 2, 87 (1902).

A. H. Cook, "The Absolute Determination of the Acceleration Due to Gravity," Metrologia 1, 84 (1965).

H. L. Curtis, Electrical Measurements: Precise Comparison of Standards and Absolute Determinations of the Units (McGraw-Hill, New York, 1937).

R. D. Cutkosky, "New NBS Measurements of the Absolute Farad and Ohm," IEEE Trans. Inst. and Meas. IM23 (4), 305 (1974).

R. L. Driscoll and R. D. Cutkosky, "Measurement of Current with the National Bureau of Standards Current Balance" [RP 2846], J. Res. Nat. Bur. Stand. (U.S.) 60, 297 (1958).

K. M. Evenson, J. S. Wells, F. R. Petersen, B. L. Danielson, and G. W. Day,, "Speed of Light from Direct Frequency and Wavelength Measurements of the Methane-Stabilized Laser," Phys. Rev. Lett. 29, 1346 (1972).

Explanatory Supplement to the Astronomical Ephemeris and the American Ephemeris and Nautical Almanac (Her Majesty's Stationery Office, London, 1961).

B. F. Field, T. F. Finnegan, and J. Toots, "Volt Maintenance at NBS via 2 e/h: A New Definition of the NBS Volt," Metrologia 9, 155 (1973).

R. T. Glazebrook, "Standards of Measurement: Their History and Development," Proc. Phys. Soc. (London) 43, Part 4 (1931).

A. Gray, Absolute Measurements in Electricity and Magnetism (Macmillan, London, 1921; reprinted by Dover, New York, 1967).

W. J. Hamer, Standard Cells: Their Construction, Maintenance, and Characteristics [Nat. Bur. Stand. (U.S.), Monogr. 84] (U.S. Govt. Printing Office, Washington, 1965).

H. Hellwig, K. M. Evenson, and D. J. Wineland, "Time, Frequency, and Physical Measurement," Phys. Today 31 (12), 23 (1978).

H. Hintenberger (ed.), Nuclear Masses and their Determination (Pergamon, New York, 1957).

S. L. Howe, NBS Time and Frequency Dissemination Services [Nat. Bur. Stand. (U.S.), Spec. Publ. 432] (U.S. Govt. Printing Office, Washington, 1976).

T. P. Kohman, J. H. E. Mattauch, and A. H. Wapstra, "C^{12} as a Basis for a Unified Scale of Nuclidic Masses and Atomic Weights," Phys. Today 12 (1), 30 (1959).

F. A. Laws, *Electrical Measurements* (McGraw-Hill, New York, 1938), 2nd ed.

W. Markowitz, R. G. Hall, L. Essen, and J. V. L. Parry, "Frequency of Cesium in Terms of Ephemeris Time," Phys. Rev. Lett. 1, 105 (1958).

C. H. Page and P. Vigoureux (eds.), *The International Bureau of Weights and Measures (1875 - 1975) [Nat. Bur. Stand. (U.S.), Spec. Publ. 420]* (U.S. Govt. Printing Office, Washington, 1975).

Precision Measurement and Calibration [Nat. Bur. Stand. (U.S.), Spec. Publ. 300] (U.S. Govt. Printing Office, Washington, 1968-), 12 volumes.

J. H. Sanders, *The Fundamental Atomic Constants* (Oxford Univ. Press, New York, 1965), 2nd ed.

D. R. Tate, "Acceleration Due to Gravity at the National Bureau of Standards," J. Res. Nat. Bur. Stand. (U.S.) 72C, 1 (1968).

B. N. Taylor, W. H. Parker, and D. N. Langenberg, "Determination of e/h, Using Macroscopic Quantum Phase Coherence in Superconductors: Implications for Quantum Electrodynamics and the Fundamental Physical Constants," Rev. Mod. Phys. 41, 375 (1969).

"The International Practical Temperature Scale of 1968," Metrologia 5, 36 (1969).

P. Vigoureux, *Units and Standards of Electromagnetism* (Springer-Verlag, New York, 1971).

J. W. T. Walsh, *Photometry* (Constable, London, 1958), 3rd ed., Chap. 6.

Standards Organizations

American National Metric Council (ANMC)
1625 Massachusetts Avenue, N.W.
Washington, D.C. 20036

A private, nonprofit organization whose purpose is to provide guidance and coordination for voluntary conversion to the metric system in the United States.

American National Standards Institute (ANSI)
1430 Broadway
New York, NY 10018

Private, nonprofit federation that promotes development of voluntary standards and represents the United States in the ISO and IEC.

International Bureau of Weights and Measures (BIPM)
Pavillon de Breteuil
F-92310 Sevres, France

Established by the Treaty of the Meter (1875), supervised by the International Committee for Weights and Measures (CIPM) under the authority of the General Conference on Weights and Measures (CGPM).

Official journal is Metrologia, published by:
Springer-Verlag New York, Inc.
Service Center Secaucus
44 Hartz Way
Secaucus, NJ 07094

International Electrotechnical Commission (IEC)
Geneva, Switzerland

International organization for engineering standards in the electrical and electronics industries. Affiliated with ISO. Represented in USA by ANSI.

International Organization for Standardization (ISO)
Geneva, Switzerland

Concerned with standardization in all fields except electricity. Represented in USA by ANSI.

International Union of Pure and Applied Physics (IUPAP)
Secretary-General:
Professor L. Kerwin
Bureau du Recteur
Universite Laval
Quebec, P.Q., Canada G1K 7P4

National Bureau of Standards (NBS) (U.S.)
Special Publications available from:
Superintendent of Documents
U.S. Government Printing Office
Washington, DC 20402

Letter Circulars available from:
U.S. Dept. of Commerce
National Bureau of Standards
Washington, DC 20234

U.S. Metric Association
Sugarloaf Star Route
Boulder, CO 80302
National, nonprofit organization founded in 1916 to promote use of the metric system in the United States. Publishes USMA Newsletter. Affiliated with the American Association for the Advancement of Science (AAAS).

Standard for Metric Practice is available from ANSI or directly from:
American Society for Testing and Materials (ASTM)
1916 Race Street
Philadelphia, PA 19103

Institute of Electrical and Electronics Engineers (IEEE)
345 East 47th St.
New York, NY 10017